The Origin of Chirality in the Molecules of Life
A Revision from Awareness to the Current Theories and Perspectives of this Unsolved Problem

The Origin of Chirality in the Molecules of Life
A Revision from Awareness to the Current Theories and Perspectives of this Unsolved Problem

Albert Guijarro and Miguel Yus
Department of Organic Chemistry, University of Alicante, Alicante, Spain

RSCPublishing

ISBN: 978-0-85404-156-5

A catalogue record for this book is available from the British Library

© Albert Guijarro and Miguel Yus, 2009

All rights reserved

Apart from fair dealing for the purposes of research for non-commercial purposes or for private study, criticism or review, as permitted under the Copyright, Designs and Patents Act 1988 and the Copyright and Related Rights Regulations 2003, this publication may not be reproduced, stored or transmitted, in any form or by any means, without the prior permission in writing of The Royal Society of Chemistry or the copyright owner, or in the case of reproduction in accordance with the terms of licences issued by the Copyright Licensing Agency in the UK, or in accordance with the terms of the licences issued by the appropriate Reproduction Rights Organization outside the UK. Enquiries concerning reproduction outside the terms stated here should be sent to The Royal Society of Chemistry at the address printed on this page.

Published by The Royal Society of Chemistry,
Thomas Graham House, Science Park, Milton Road,
Cambridge CB4 0WF, UK

Registered Charity Number 207890

For further information see our web site at www.rsc.org

Preface

"*The story of a theory's failure often strikes readers as sad and unsatisfying. Since science thrives on self-correction, we who practice this most challenging of human arts do not share such a feeling. We may be unhappy if a favored hypothesis loses or chagrined if theories that we proposed prove inadequate. But refutation almost always contains positive lessons that overwhelm disappointment, even when no new and comprehensive theory has yet filled the void.*"

Stephen Jay Gould

There can have been very few occasions on which an unresolved issue in science such as *The Origin of Chirality in the Molecules of Life* has been the subject of such a large number of approaches, nor that the theories intended to explain it have arisen from such widely different perspectives. Ranging from sophisticated theories in quantum physics and the complexity of evolutionary biochemistry, to the simple trivialization of the problem, this topic leaves hardly anybody impassive. Even after more than 150 years of discussion the issue remains a provocative open question.

Attempting to forecast the future in science can be fun, but more often than not it is a hopeless cause. Very occasionally a scientist of outstanding intelligence appears, and Louis Pasteur was one of these. He arrived at the conclusion that a *dissymmetric force* in Nature was in some way responsible for biomolecular handedness. From this theory he initiated the search for *deterministic theories* which might explain the contemporary observation of homogeneous chirality in biomolecules—but he never saw the fruits of his search. Pasteur's original concept of a dissymmetric force in the Universe was in essence correct, however. Although he could hardly have imagined its fundamental nature and extent, it was in some respects a premonition of the discovery of the weak force a century later. The weak force is a permanent chiral bias acting everywhere and at all times in the Universe. It is thought—although this has not so far been established quantitatively—to be responsible for what is probably the largest asymmetry occurring in the Universe, the overabundance of matter rather than antimatter, or charge asymmetry. It also determines the arrow of time, signifying the direction of the time–flow in particle physics, and it allows a preference for an absolute chirality, establishing an energy difference between enantiomers.

At our current level of knowledge and our ability to interpret the facts, the origin of these asymmetries is tied in with the forces existing immediately after the Big Bang, and specifically with the weak force. Being left-handed in character, the weak force imprinted its sign on the evolving Universe. However, when extrapolated to the molecular level—of particular relevance to us here—the weak force is unable to provide a complete answer to the origins of biomolecular homochirality. It has still to be unequivocally established whether the connection between the preference of the weak force for one enantiomer, and the homochirality displayed by living organisms, exists or not, and if the one is the cause of the other.

"One never notices what has been done; one can only see what remains to be done."

Marie Curie

Dozens of alternative deterministic theories have been postulated in the past, many of them resting on solid ground but in every case unsupported by clear experimental evidence. Theories that rely on long or even evolutionary time-scales, mainly due to the smallness of the suggested chiral bias, have little chance of ever being tested experimentally. In other cases, rather like fashionwear, some of the better older theories have drifted out of favor and acceptance and have become unfashionable, overshadowed by theories which although more recent are not necessarily better. Re-examination of these ideas, faded with time but still good in shape, has been felt worthwhile and has been one of the goals of this work.

"A philosopher once said, 'It is necessary for the very existence of science that the same conditions always produce the same results.' Well, they don't!"

Richard P. Feynman

This quote could be taken as an introduction to an alternative group of theories considered in this book, those based on *chance mechanisms*. In chance theories there is no requirement for Pasteur's dissymmetric force or chiral influence. Chance mechanisms inherently produce a random outcome to ultimate macroscopic chirality. In any event, an amplification of an initial random chiral fluctuation at the molecular level seems to be the underlying mechanism, the robustness of this amplification determining the feasibility of the process.

Preface

The replication of a polynucleotide of randomly generated chirality is one obvious choice for such an autocatalytic growth mechanism. On the other hand, a similar role could be provided by certain polypeptides, or other as yet undetermined macromolecules of prebiotic relevance. Within these theories, the ultimate sign displayed by molecules in the biosphere has developed by chance—and it is certainly untestable. But some mechanisms should still explain the randomness of the outcome and possibly reproduce this effect.

"It seems plain and self-evident, yet it needs to be said: the isolated knowledge obtained by a group of specialists of a narrow field has in itself no value whatsoever, but only in its synthesis with all the rest of knowledge and only inasmuch as it really contributes in this synthesis towards answering the demand, 'who are we?'"

Erwing Schrödinger

Some of the theories discussed in this book could be discarded if we were capable of addressing another old but still unanswered question: is life a common feature of the Universe? Would it be reasonable to expect life to have emerged in other planets under conditions similar to those on Earth? Research efforts on the origin of chirality in the molecules of life run parallel to those involved in the search for the origins of life, and they could retrieve important information from the same sources. Exobiology is one of these. Undoubtedly the search for molecular life or its vestiges in other planets is of the upmost importance. More accessible, although still a difficult task, is the search for traces of fossil life in Archean sediments; of special significance (and difficulty) would be to rescue any traces of organic matter, possibly prebiotic, that may somehow have escaped destruction during geological processes. Replication of plausible prebiotic reactions in the laboratory and developing them to the point of giving them self-replicative and evolutionary functions akin to living matter is another of these fronts, at the moment perhaps the most accessible.

If life at any molecular level is in fact such a highly exceptional occurrence, every theory has a chance of being correct, including those which seem at first sight most improbable. On the other hand, if life is indeed a common feature—and this is perhaps the more widely accepted view—theories based on a remote combination of factors become less convincing. Clues to this puzzle are likely to emerge from studies in widely different fields, including geology, biology, chemistry, physics and astronomy, and will involve the expertise of a broad interdisciplinary array of scientists. Important pieces of information will arise from observations at the extremes of two of the frontiers of science, outer space and the subatomic world. Observations on distant planets and galaxies, and even the actual sampling of celestial objects outside the solar system, are projects currently in hand. At the opposite end of the spectrum experiments are planned which will reveal the elemental properties of matter beyond the framework of current physics, such as symmetry and interactions between the fundamental forces of Nature. We are at present only beginning to appreciate the levels at which fundamental forces might operate, and we can be confident that these efforts are about to bear fruit.

Finally, we wish to pay tribute to the protagonists of this work: those scientists from all fields of science who have contributed in one way or another to its shaping. Arising from their stimulating teaching and the heated discussion which have often been generated, we have made an attempt to collect the essentials of the theme as it has evolved over a period of many years. We apologize for any omissions which may come to light. Unfortunately these are almost inevitable in compiling such a vast and interdisciplinary review and condensing it to a concise format. Finally, we extend our special thanks to those with whom we have shared stimulating discussions, and who have enlightened our thinking. We are also grateful to all those who have played a part in the editing and production process.

Contents

Chapter 1 Introduction and Historical Background

    1.1 Introduction   1
    1.2 The Contribution of Pasteur   1
           1.2.1 Quartz   2
           1.2.2 Tartaric Acid and the Tartrates   3
    1.3 The Nature of the Problem   5
    References   5

Chapter 2 Theories of the Origin of Biomolecular Homochirality

    2.1 Introduction   6
    2.2 Chance Theories   7
           2.2.1 Amplification of Tiny Stochastic Imbalances   8
           2.2.2 Polymerizations Above a Critical Chain Length: Macromolecules   10
    2.3 Deterministic Theories   17
    References   18

Chapter 3 The Concept of Chirality

    3.1 Introduction   21
    3.2 The Basic Symmetry Operations   22
           3.2.1 Space Inversion: The Parity Operator, $\hat{P}$ 22
           3.2.2 The Time Reversion: The Time Reversal Operator, $\hat{T}$ 23
           3.2.3 The Charge Conjugation: The Charge Conjugation Operator, $\hat{C}$ 24

The Origin of Chirality in the Molecules of Life
Albert Guijarro and Miguel Yus
© Albert Guijarro and Miguel Yus, 2009
Published by the Royal Society of Chemistry, www.rsc.org

3.3	The Failure of the Experiments of Pasteur, Curie and Others	25
	3.3.1 Absence of Chiral Field	25
	3.3.2 Experiments Under False Chiral Influence	25
	3.3.3 Experiments Under True Chiral Influence	26
3.4	Expected Effect of True and False Chirality on Molecules	28
	References	30

Chapter 4 Chiral Physical Forces

4.1	Background	31
4.2	The Weak Interaction and the Parity Violation	31
	4.2.1 The Fundamental Interactions: Gravitational, Weak Interaction, Electromagnetic and Strong Interaction	31
	4.2.2 The β-Decay	34
	4.2.3 Parity Violation	35
	4.2.4 Unification of Forces	38
	4.2.5 Quantification of the Parity Violation in Molecules: ΔE_{pv}	41
	4.2.6 β-Radiolysis	46
4.3	Asymmetric Photolysis and Photosynthesis	48
	4.3.1 Circularly Polarized Light: Circular Dichroism	48
	4.3.2 The Magnetochiral Effect	57
4.4	Fluid Dynamics: Vortex Motion	59
	References	65

Chapter 5 Mechanisms of Amplification

5.1	Background	72
5.2	Autocatalysis	72
	5.2.1 The Frank Model	72
	5.2.2 Theoretical Models Derived from Frank's Original Model[6]	75
	5.2.3 Autocatalytic Amplification of the Weak Force	77
	5.2.4 Asymmetric Autocatalytic Reactions	78
	5.2.5 Self-replication	82
5.3	Amplification of Chirality in Other Chemical Reactions	84
	5.3.1 Nonlinear Effects	84
	5.3.2 Amplification of Chirality by Cooperative Forces: Growing Polymers and Supramolecular Assemblies	86
5.4	The Yamagata Cumulative Mechanism	88

Contents

5.5	The Salam Phase Transition	91
5.6	Amplification of Scalemic Compounds: Eutectic Mixtures	91
	5.6.1 Solubility Properties	93
	5.6.2 Sublimation Properties	97
5.7	Amplification of Chirality in Serine Octamers	98
	5.7.1 Homochiral Preference and Chiral Amplification	98
	5.7.2 Chemistry of Serine Octamers	99
	5.7.3 Sublimation Experiments	102
References		102

Chapter 6   Spontaneous Symmetry Breaking

6.1	Introduction	108
6.2	Spontaneous Symmetry Breaking in Crystallization	110
References		113

Chapter 7   Outside Earth: Meteorites and Comets

7.1	Introduction	115
7.2	Meteorites	115
7.3	Comets	120
References		123

Chapter 8   Other Local Deterministic Theories

8.1	Introduction	125
8.2	Chiral Crystals and Faces on Crystals	125
	8.2.1 Quartz	125
	8.2.2 Calcite, Gypsum, Clay Minerals and Others	129
	8.2.3 Organic Crystals: Glycine	132
8.3	Two-Dimensional Chirality	132
References		135

Chapter 9   Intrinsic Asymmetry of the Universe: The Arrow of Space–Time and the Unequal Occurrence of Matter and Antimatter

References	145

Subject Index   146

CHAPTER 1
Introduction and Historical Background

1.1 Introduction

Chirality, or handedness, in molecules related to living organisms has fascinated scientists ever since the phenomenon was first observed, and it remains a fundamental question still not fully explained. Of the two possible series of enantiomeric molecules, why did Nature choose the L-amino acids and D-sugars when creating the structures of life? Why not the other way round? Indeed, why not both, which is at first sight the most likely chemical outcome?

For some scientists this is an encrypted clue provided by Nature to unveil its origins. For others it is much less than that, merely a matter of chance. We will describe in this book the most relevant pieces of information gathered by scientists over the past 150 years concerning this issue. Along the way we will explore some of the basic principles of the laws of Nature, principles which govern all the processes in the Universe, some of them so profound that they approach the limits of scientific knowledge, and which have influenced the nature of the Universe since its origin.

1.2 The Contribution of Pasteur

We are indebted to Louis Pasteur for the first theory on the origins of biomolecular homochirality—in fact we owe not only the initial theory to him but also our awareness of the problem itself. Most chemists are familiar with Pasteur's work on the resolution of tartaric acid (later known as racemic acid) into its enantiomers, a procedure which established the foundations of molecular stereochemistry (Figure 1.1).[1] The historic chain of events which led in the middle of the nineteenth century to the development of molecular stereochemistry had its

Figure 1.1 Louis Pasteur

origins in the observation of chirality in compounds obtained from living organisms. Correlation of their crystallographic properties with those found in mineral samples—mineralogy being the better developed field at the time—provided the necessary understanding for this.

1.2.1 Quartz

Quartz had an important role to play. At the beginning of the nineteenth century, in 1801, the crystallographer R. H. Haüy observed that the apparent hexagonal symmetry of quartz crystals was in fact reduced in most cases by the presence of small faces called hemihedral facets (hemihedral meaning that only half the faces required for complete symmetry were exhibited) at alternate corners of the crystal. The presence of these hemihedral facets has a profound effect on symmetry. It eliminates the centre and planes of symmetry of the basic holohedral (holohedral indicating that it has the highest symmetry) hexagonal crystal, and gives rise to two non-superimposable mirror image forms of quartz, both chiral and enantiomorphic, which can be recognized by their outward aspect (further details are given in Section 8.2, *Chiral Crystals and Faces on Crystals*). Around the same time, the discovery of optical activity—a necessary tool in the study of chirality on the molecular scale—was attributed

Introduction and Historical Background

to the mathematician-physician, F. Arago in the early nineteenth century (1811), and it is equally related to the mineral quartz, since this was the first material in which optical rotation was observed.[2] Soon afterwards the physicist, J.-B. Biot, discovered that natural quartz existed in two forms which rotated the plane of polarization in opposite directions, and he also established a linear relationship between the magnitude of the angle of rotation and the thickness of the slice of quartz (1812). In addition, he introduced and refined the polarimeter as a scientific tool. The two forms of quartz which Biot found to rotate in opposite senses in his polarimeter were subsequently identified by J. W. F. Herschel (1822) as the two hemihedral forms.

1.2.2 Tartaric Acid and the Tartrates

The wave theory of light was gaining acceptance at the time. Defended among others by the physicist A. J. Fresnel, the theory of transverse waves led to the conclusion around 1824 that linearly polarized light might be considered to be the superimposition of left- and right-circular polarized light. From this, it followed that the optical rotation was the consequence of the different refractive index of the two beams when passing through a chiral medium. Biot noticed that the effect of optical rotation in the plane of polarized light was not specific to crystals but was also found with certain natural products in the liquid state, including turpentine, aqueous solutions of sugar or tartaric acid (in 1832), and even vapors of such substances where they were volatile.

Later in the century (1843) Biot gave an intriguing account of certain anomalous relationships between two isomeric substances of formula $C_4H_6O_6$, the naturally occurring (+)-tartaric acid and the optically inactive paratartaric acid, in relation to the law of isomorphism, discovered earlier by the mineralogist, E. Mitscherlich. This brought Louis Pasteur on to the scene, around 1847–48. Mitscherlich had compared the crystal forms of the corresponding salts of the two acids and found that they differed in crystal morphology, those obtained from (+)-tartaric acid being hemihedral and those derived from paratartaric acid holohedral racemic crystals (using current terminology), except in one instance. This was the case of sodium ammonium tartrate and paratartrate, in which the crystals appeared identical, and these salts displayed hemihedral morphology in both cases.

Pasteur decided the latter case required further study. Witnessed by Biot, Pasteur worked with tartaric acid, which Biot had shown to be optically active, and with paratartaric acid, which was chemically identical but optically inactive, and prepared crystals of the corresponding sodium ammonium salts. He showed that although both salts were indeed hemihedral, in the (+)-tartrate the hemihedral facets were all facing in the same direction, whereas in the paratartrate there were equal amounts of crystals with hemihedral facets having either this orientation or the opposite, forming a conglomerate of enantiomorphous crystals (for the definition of a conglomerate, see Section 5.6, *Amplification of Scalemic Compounds: Eutectic Mixtures*).

Figure 1.2 illustrates two actual enantiomorphous crystals of sodium ammonium tartrate, similar to those obtained by Pasteur in his work with racemic acid. In Figure 1.3b drawings of the two enantiomorphic crystals are shown, taken from Pasteur's original notes. These crystals are enantiomorphous, since they are mirror images and are not superimposable. Pasteur performed the first enantiomeric resolution of the crystals using tweezers and a magnifying glass.[3] At the time he was aware of the work of Herschel, who had reported more than 20 years previously that left-handed quartz crystals were levorotatory whereas the corresponding right-handed crystals were dextrorotatory, *i.e.* enantiomorphous quartz crystals rotated the plane of polarized light in opposite directions (Figure 1.3b, and Section 3.2, *Chiral Crystals and Faces on Crystals*).[4]

Figure 1.2 Large crystals of sodium ammonium tartrate prepared by the seeding method. Left, (−)-enantiomer; right, (+)-enantiomer. (From Reference 5, with authorization.)

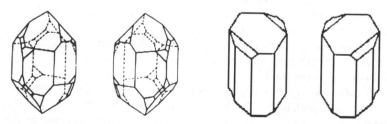

a) Left-handed and right-handed quartz crystals (rotations in solid phase) b) (−)- And (+)-sodium ammonium tartrate (rotations in solution)

Figure 1.3 (*a*) Enantiomorphic crystals of quartz from nature; (*b*) Enantiomophic crystals of sodium ammonium tartrate obtained from racemic tartaric acid.

1.3 The Nature of the Problem

Crystals have a highly orderly structure. It seemed reasonable that chiral solids such as quartz or tartrate salts should be optically active. Unlike quartz, which is insoluble, tartrates are very soluble in water and remain optically active in solution, as demonstrated earlier by Biot. Since liquid phases are not organized structures, chirality is therefore a property of the molecules themselves. One of the two enantiomorphous crystals of the racemate yielded (+)-tartaric acid after acid treatment, identical to the tartaric acid deposited by maturing wines. On the other hand, the optically inactive racemic acid, consisting of a mixture of equimolecular amounts of (+)- and non-natural (−)-tartaric acid, was obtained for example by synthesis (among other sources), and in common with chemical syntheses at the time this was a racemic synthesis.

It is at this point that the fundamental question has to be asked: why does Nature display a preference for one of these two apparently equivalent molecules? Moreover, why is there a preference of this kind with most organic molecules in living organisms? Pasteur proposed at that time the first theory of the origins of biomolecular homochirality: the existence of chiral, or as he described them, *dissymmetric* forces in Nature.[6] He dedicated in vain the remainder of his career as a chemist to the search for such dissymmetric forces in Nature. Perhaps his frustration in this area of chemistry helped to divert his efforts towards other fields of science. Indeed, his studies in microbiology were to provide the foundations of pathology, a keystone of medicine itself. Throughout this book we will see that, while he was unable to find these chiral forces, and that he was certainly mistaken interpreting some physical aspects concerning their chirality, his concept of an underlying asymmetric tendency in Nature was essentially correct.

References

1. A detailed chronological description of the historic events and the numerous scientists contributing to this critical period of the science that gave birth to the stereochemistry can be found in: (a) S. F. Mason, in *Chirality in Natural and Applied Science*, ed. W. J. Lough and I. W. Wainer, Blackwell Science, Oxford, 2002, 1–21; (b) L. D. Barron, *Molecular Light Scattering and Optical Activity*, 2nd edn, Cambridge University Press, Cambridge, 2004, 1–52.
2. F. Arago, *Mém. Inst.*, 1811, **1**, 93–134.
3. L. Pasteur, *C. R. Acad. Sci. Paris*, 1848, **26**, 535–539.
4. J. W. F. Herschel, *Trans. Cambridge Philos. Soc.*, 1822, **1**, 43–52.
5. I. Tobe, *Mendeleev Commun.*, 2003, 93–94.
6. L. Pasteur, *Leçons de Chimie Professées en 1860*, Lib. Hachette, Paris, 1861, 1–48.

CHAPTER 2
Theories of the Origin of Biomolecular Homochirality

2.1 Introduction

Experimental observations accumulated over the years have provided evidence leading—either tightly or loosely—to the various theories of the origins of biomolecular homochirality. They form a jigsaw of which some pieces are missing—perhaps in the form of definitive evidence—but this has not prevented preliminary interpretations of this puzzle being offered. And there have been many. Investigators have proposed a variety of hypotheses, but evidence in favor of each of them is at best fragmentary. Until a deeper level of understanding of the problem is reached and one theory prevails, there is a need to classify the various theories which have been proposed, possibly using different perspectives.[1]

The first question to be asked in such a classification is whether there was in fact a cause, a specific chiral bias which provoked the mirror-symmetry breaking observed with biomolecules. If the answer to this question is negative, we are dealing with theories based on *chance* mechanisms—chance in the sense of randomness—and the grounds for this will be explained in the section following. If the answer is positive, then there is a relationship between cause and effect. In this case the observed mirror-symmetry breaking is a consequence of an earlier chiral influence, even if this is on a minuscule scale. This is consistent with the philosophical proposition of determinism, and we shall discuss it in terms of *deterministic* mechanisms and theories. These, to which most of this book is devoted, given their prolific nature, can be further divided into *local* deterministic and *universal* deterministic. They are local deterministic if the initial chiral influence existed in a specific given location (local in space), or over a limited period of time (local in time), but averages zero over large enough areas of observation or long enough periods of time. On the other hand, they are described as universal deterministic if there is a permanent, inexorable

Theories of the Origin of Biomolecular Homochirality

Scheme 2.1 Classification of the various theories of the origins of biomolecular homochirality.

chiral influence, regardless of its strength (Scheme 2.1), at the time the chiral selection occurred.

Whereas in the first group, those concerning chance theories, one or more appropriate mechanisms should explain the randomness of the outcome, the ultimate chiral sign actually observed in the biosphere is certainly untestable, *i.e.*, homochirality has developed simply by chance. Conversely, deterministic theories can be subjected to experimental confirmation, since if there was in fact a chiral influence that imposed its influence this should in principle be reproducible. This classification is not without some degree of overlap. This is easily observed within the subgroup of deterministic theories, for example (a) local deterministic or regional—the synthesis of prebiotic molecules on a chiral crystal of (+)-quartz,[2] and (b) universal deterministic mechanisms—the result of the weak force.[3] Local mechanisms always rely on a chiral manifestation which has its enantiomeric counterpart elsewhere, for example on a (−)-quartz crystal, which occurs with the same probability as its enantiomers, according to the latest studies,[4] or at some other time, such as small circularly polarization of light at dawn, of opposite sign to that at dusk.[5] From this point of view, deterministic local also has an element of a chance mechanism, represented in Scheme 2.1.

Among other classifications, the so-called biotic *vs* abiotic theories regard the origin of life as the ruling criterion, which could have taken place either before the enantiodiscrimination step (biotic theories) or afterwards (abiotic theories). Biotic theories, which entail coexistence of *dextro*- and *levo*-organisms followed by the extinction of one of them, have lost their appeal for most scientists. It could be tempting to invoke the occurrence of some D-amino acids in primitive organisms as a remnant of a racemic life parallel to ours, a primeval enantiolife which could have originated with the beginnings of evolution. This does not seem to be the case. Although a few unnatural D-amino acids are found in bacterial cell walls and other secondary metabolites,[6] these are synthesized from the L-form, which is the only form coded by bacterial DNA.[7]

2.2 Chance Theories

At first sight, chance mechanisms are somewhat counterintuitive to chemists. This is in part because chance mechanisms rarely have a role in conventional chemistry due to the statistical behavior of the samples involved, which consist of a very large number of molecules, often approaching the scale of Avogadro's number. We are accustomed to reproducible experiments every time we run them. The opposite would be the macroscopic manifestation of a single microscopic stochastic event, which would give random results every time we ran it. The scenarios of spontaneous symmetry breaking can be included in this category (Chapter 6, *Spontaneous Symmetry Breaking*) and are mainly associated with physical phenomena. Chemical reactions of this type are very much less common, in fact they were purely theoretical until very recently, with the advent of Soai's reaction (Section 5.2.4, *Asymmetric Autocatalytic Reactions*).

2.2.1 Amplification of Tiny Stochastic Imbalances

In principle, the evolution of biological homochirality could be theoretically explained by a model in which a tiny imbalance of one enantiomer was exaggerated, or *amplified*, by autocatalytic reactions. A chemical reaction is autocatalytic if the reaction product is itself the catalyst for that reaction. This concept is discussed in Section 5.2 under the heading *Autocatalysis*, and is mechanistically related to the well-known Frank model.[8] However, concerning the minute imbalance, this must by definition be merely stochastic in origin. To clarify this point, some features of the racemic state will be summarized.[9]

2.2.1.1 *The Racemic State*

In general, chiral molecules are obtained in the racemic state when synthesized from achiral reagents, for example inorganic or small organic starting materials. Racemism is, however, an intrinsically macroscopic attribute. A statistical description of the racemic state down to the individual molecule is, however, of interest for a variety of reasons. In principle the racemic state can be described by a binomial distribution.[10] In its broad definition binomial distribution gives the discrete probability distribution $P_p(n, N)$ of obtaining exactly n successes out of N Bernoulli trials, where the result of each Bernoulli trial occurs with probability p (and therefore the opposite occurs with probability $1-p$), represented by Equation (2.1). This treatment is subject to the constraint of the strict independency of each Bernoulli trial in regard to what has happened before. Tossing a coin is a good analogy, where the result of each flip, either head or tail, is independent of the previous history of results and has a probability $p = 1/2$. Translated to our racemic synthesis, this implies that are dealing with an irreversible synthesis, where the reaction products (D-and L-molecules) do not have any effect on the previous synthetic steps. A drawing of the probability distribution for this racemic synthesis is shown in Figure 2.1. As expected, the *mean value* of this distribution is $\mu = Np = N/2$, while the *standard deviation* is $\sigma = \sqrt{Np(1-p)} = \sqrt{N}/2$, in agreement with the properties of the binomial distribution. If the enantiomeric molecules D and L are synthesized with exactly

Theories of the Origin of Biomolecular Homochirality

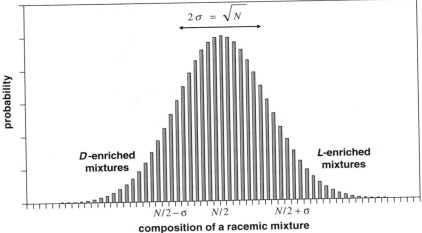

Figure 2.1 The statistical distribution of the composition of racemic mixtures can be derived from the binomial distribution, which is characterized by its mean value, $\mu = N/2$ and the standard deviation, $\sigma = \sqrt{N}/2$. A racemic mixture is seldom equimolecular, the magnitude of its dispersion being given by σ. In plain numbers, about 68% of racemic mixtures (area in red) are within one standard deviation from the mean value, while for the remaining 32% (in blue) the deviation from $N/2$ is greater.

the same probability, *i.e.*, $p = 1/2$, the binomial distribution gives the discrete probability distribution $P(N/2, N)$ of having exactly $N/2$ molecules of one kind (*e.g.*, of D-configuration) and $N/2$ of the other (disregarding for simplicity the case in which N is odd, the effect of which vanishes for N large, in which case there are two *modes* instead of one, symmetrically centered around $N/2$). This probability is expressed by Equation (2.2), where the combinatorial numbers have been calculated using the Stirling approximation, Equation (2.3).

The result is self-explanatory. The probability of having an exact racemic mixture tends to zero with the square root of N, which, for one mol of a racemic compound ($N = N_A$), amounts to the insignificantly small probability of $P(N_A/2, N_A) \approx 10^{-12}$. So, although the most likely mixture or mode is $(N/2)_D + (N/2)_L$ since it lies on top of the bell (Figure 2.1), we rarely have truly racemic mixtures. This assertion gains certain relevance when amplification mechanisms are discussed.

For N large, the binomial distribution approaches a normal (Gaussian) distribution and converges as N is taken to infinity, Equation (2.4). In Equation (2.4) the sum over discrete probabilities $P(n, N)$ between $n = a$ and $n = b$ is substituted by an integral of a continuous probability function over the same limits, $x = a$ to b, which takes the form of the normal distribution in the $N \to \infty$ limit.[11] This allows the probability function to be integrated, Figure 2.1. In this the area in red is within one standard deviation from the mean ($N/2$). It covers about 68% of values drawn from a standard normal distribution. Also, from the properties of the normal distribution, about 95% of the values are within two standard deviations and about 99.7% lie within three standard deviations. As seen previously, the standard deviation is for one mol, $\sigma = \sqrt{N_A}/2 \approx 3.9 \cdot 10^{11}$, which gives an idea of the magnitude of the stochastic dispersion in the number of molecules, from the exact equimolecular number of D- and L-molecules in one mol in the racemic state. This dispersion from the mean value is certainly well below the detection limit of any current analytical technique, but it can be advantageously compared with other proposed minor chiral effects, such as the parity-violating weak force (Section 4.2.5, *Quantification of the Parity Violation in Molecules:* ΔE_{pv}).

$$P_p(n, N) = \binom{N}{n}(p)^n(1-p)^{N-n} \tag{2.1}$$

$$P(N/2, N) = \binom{N}{N/2}\left(\frac{1}{2}\right)^{N/2}_L\left(\frac{1}{2}\right)^{N/2}_D \approx \sqrt{\frac{2}{\pi N}} \tag{2.2}$$

$$N! \approx \sqrt{2\pi N}\left(\frac{N}{e}\right)^N \tag{2.3}$$

$$\lim_{N \to \infty} \sum_{n=a}^{b} \binom{N}{n}\left(\frac{1}{2}\right)^N = \frac{1}{\sigma\sqrt{2\pi}} \int_a^b e^{\frac{(x-\mu)^2}{2\sigma^2}} dx, \quad \text{with} \quad \mu = N/2, \sigma = \sqrt{N}/2 \tag{2.4}$$

As mentioned above, this scenario is rigorously valid only for an irreversible racemic synthesis, or, in other words, for a kinetically driven racemic synthesis, or under kinetic control, where there is no way back for the reaction products (D- and L-molecules) to the reagents. The opposite situation is the thermodynamically driven racemic synthesis, or under thermodynamic control, where reversibility allows for spontaneous correction of the stochastic divergences towards the exact equimolecular mixture of enantiomers, which is the one with the highest entropic content. This reversible scenario does not fully comply with the requirement to be independent of previous history in a series of Bernoulli trials, from which the binomial distribution is derived. Indeed, at thermodynamic equilibrium, in the absence of any chiral influence the mixture tends to be strictly equimolecular in enantiomers, giving allowance to the existence of stochastic fluctuations, which tend to be time-averaged to zero.

2.2.2 Polymerizations Above a Critical Chain Length: Macromolecules

Another approach which has evolved into a chance theory was formulated by analyzing the statistics of the polymerization processes, in either synthetic or biomolecular polymers. It was during the time polymer sciences were being developed[12] that attention was drawn to the fact that an atactic vinyl polymer is actually a mixture of chiral chains, provided the degree of polymerization is high enough.[13]

To illustrate this, let us consider a range of samples of commercial polystyrene ($M_w \cong 2\,000\,000$, $900\,000$, $400\,000$, $200\,000$, $90\,000$ and $35\,000$; $DP \cong 19\,000$, 8600, 3800, 1900, 860 and 330), or of PVC ($M_w \cong 225\,000$, $151\,000$, $101\,000$, $66\,000$, $36\,500$ and $11\,200$; $DP \cong 3600$, 2400, 1600, 1000, 580 and 180) (M_w = weight average molecular weight; DP = degree of polymerization, or length (n) of a representative chain of molecular weight = M_w).[14] For most industrial applications, chain lengths in the thousands or tens of thousands are typical. Let us evaluate now the number of chiral chains for a given chain length. (The calculation of the numbers is not critical for an adequate understanding of the ultimate objective, but it is instructive for the analysis of complex mixtures). In an atactic polymer, for a given chain length n, the total number of possible configurational sequences is $N_{total} = 2^n$. Ignoring the effect of the chain end-groups (initiation/termination), which in high molecular weight polymers are structurally unimportant, most of these sequences come in pairs of chemically equivalent chains. We shall use N for the calculated sequences and N for the actual chemically nonequivalent chains. *Meso* forms can be calculated by considering: (a) for n even, all the permutations with repetition of half of a chain, affording $^{even}N_{meso} = 2^{n/2}$; since they come in pairs of equivalent chains, half of this is the actual number of *meso* forms, $^{even}N_{meso} = 2^{n/2}/2 = 2^{(n-2)/2}$; (b) for n odd, $^{odd}N_{meso} = 2 \times 2^{(n-1)/2} = 2^{(n+1)/2}$, where the central stereogenic carbon (or pseudoasymmetric r or s) requires multiplying by 2 all the permutations of the half chain of $(n-1)/2$ length (Figure 2.2). These sequences also come in pairs of equivalent chains, so the actual number of *meso* forms is $^{odd}N_{meso} = 2^{(n-1)/2}$. *Meso* chains are not the only chains displaying certain symmetry. Conceptually, each sequence in a half-chain has two possible arrangements in the second half of the complete chain. The second half-chain can be connected with identical configuration, affording *meso* sequences (*e.g.*, RSRS, or RSrRS/RSsRS for n odd), or with opposite configuration, affording *palindromic* sequences, *e.g.*, RSSR (n even), or RSSSR ≡ RSRSR ≡ RSASR, where A stands for a non-stereogenic carbon center, which occurs for n odd. Consequently, $N_{meso} = N_{palin}$. *Meso* forms are achiral, exist in equivalent pairs (RSRS ≡ SRSR), and should not be included in the overall counting. Palindromic forms are chiral and must be counted in, but their sequences come not in pairs, but alone for n even (*e.g.*, RSSR, which has no other chemically equivalent sequence), and are only in pairs for n odd (*e.g.*, RSASR, due to $A = $ "R" or "S," making both sequences equivalent). Therefore $^{even}N_{palin} = {^{even}N_{palin}}/2^{n/2}$, and $^{odd}N_{palin} = {^{odd}N_{palin}}/2 = 2^{(n-1)/2}$. This is the only difference between even and odd chain counting. Finally, by far the most numerous group of chiral chains corresponds to the remaining fully asymmetric chiral sequences up 2^n, which also appear in pairs of equivalent chains (*e.g.*, sequences such as ···RRRSRSR··· ≡ ··· RSRSRRR··· correspond to one and the same molecule) and its number is $N_{asymm\ chiral} = (N_{total} - N_{meso} - N_{palin})/2$. Counting the total chiral asymmetric and chiral palindromic sequences gives the total number of chiral chains,

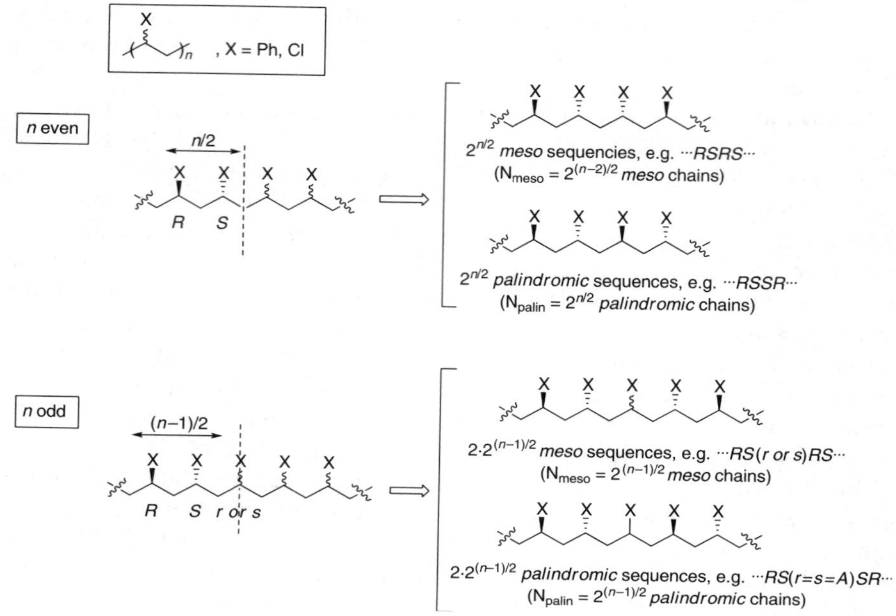

Figure 2.2 In an atactic vinyl polymer, for a given chain length n there are theoretically a total of $N_{\text{total}} = 2^n$ random sequences of diastereomers. Ignoring the effect of the end groups, the *meso* chains (achiral) and palindromic chains (chiral) can be totalled, the exact number of chiral chains being given by the expression $N_{\text{chiral}} = (N_{\text{total}} - N_{\text{meso}} - N_{\text{palin}})/2 + (N_{\text{palin (for } n \text{ even})}$, or $N_{\text{palin}}/2$ (for n odd)). N_{chiral} differs slightly depending on whether n is even or odd, but is of the order $N_{\text{chiral}} \approx 2^{n-1}$ for n large.

Equation (2.5):

a) for n even :
$$N_{\text{chiral}} = \frac{N_{\text{total}} - N_{\text{meso}} - N_{\text{palin}}}{2} + N_{\text{palin}}$$
$$= \frac{2^n - 2^{n/2} - 2^{n/2}}{2} + 2^{n/2} = 2^{n-1}$$

b) for n odd :
$$N_{\text{chiral}} = \frac{N_{\text{total}} - N_{\text{meso}} - N_{\text{palin}}}{2} + \frac{N_{\text{palin}}}{2}$$
$$= \frac{2^n - 2^{(n+1)/2}}{2} = 2^{n-1} - 2^{(n-1)/2} \qquad (2.5)$$

These expressions can be further simplified for n large. For increasing chain lengths (n), both expressions in Equation (2.5) converge very rapidly to $N_{\text{chiral}} \cong 2^{n-1}$. Indeed, the total number of diastereomers (*chiral* + *meso*, n even or odd) is also very well approximated by $N_{\text{total}} \cong 2^{n-1}$, since the number of the

Theories of the Origin of Biomolecular Homochirality

meso chains is small compared to the chiral chains for n large. The meaning of fast convergence is better appreciated considering that, for $n \geq 78$ (n not so large), the error in not considering the minor $2^{(n-1)/2}$ term is already smaller than, for example, the standard deviation in the mean composition of a mol of a racemic compound ($\sigma = \sqrt{N_A}/2$).

At the scale of one mol of monomer (*e.g.*, $PhCH=CH_2$ or $ClCH=CH_2$), the maximum number of available chains of length n is $N_{avail} = N_A/n$, where N_A is Avogadro's number. The actual number of available chains of length n is actually much smaller, since no polymerization technique gives chains of unique length but a rather distribution of chain lengths. On the other hand, the number of possible chiral chains is, as we have seen, $N_{chiral} \cong 2^{n-1}$, which increases exponentially with n. As the degree of polymerization (n) increases, these functions, monotonically decreasing and increasing respectively, intercept near $n = 74$ (analytic intersection at $n = 73.79$) (Figure 2.3). If n is to some extent smaller than the intersection point the polymer will be racemic. However, for n near to or larger than the intersection point, the number of possible chiral chains (2^{n-1}) will be well above the number of available polymeric chains (N_A/n), and the sample will merely consist of a random representation of

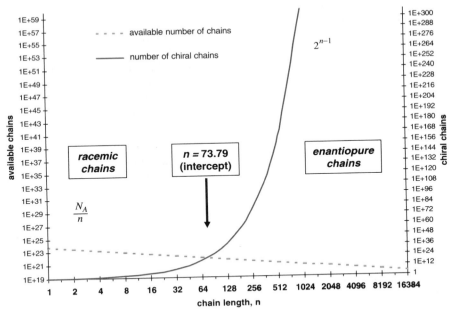

Figure 2.3 In the atactic polymerization of one mol of vinyl monomer, the maximum number of available chains of length n is $\approx N_A/n$ while the number of possible chiral chains is $\approx 2^{n-1}$. Both functions intercept near $n = 74$. If n is somewhat smaller than the intersection point, the polymer will be racemic, while for greater degrees of polymerization the polymer will be a mixture of random enantiopure chiral chains. All the scales are logarithmic.

enantiomerically pure diastereomers. Numerically, for the above mentioned polymers, e.g., PVC $M_w \cong 11200$ ($DP \cong 180$), only one out of 2×10^{32} possible chains are actually present in one mol of polymerized monomer. For polystyrene (PS) $M_w \cong 35000$ ($DP \cong 330$), the ratio is 1: 6×10^{77}. The number of possible chains per each actual chain in a polymer of $DP = 404$ is already larger than a googol (10^{100}). For PVC $M_w \cong 36500$ ($DP \cong 580$), the ratio is only $1:10^{153}$, for PS $M_w \cong 90000$ ($DP \cong 860$), it is $1:10^{237}$, and for PVC $M_w \cong 66000$ ($DP \cong 1000$) it is $1:10^{280}$. For polymers with $DP > 1000$, the number is off-scale by normal computation. For all of them, the chances of finding a single racemic chain, i.e., a single polymer chain compensated with its enantiomer, is virtually zero.

These facts, when connected to the evolutionary scenario, emerge as an important theory which significantly assists the understanding of the mirror-symmetry breaking seen in the biomolecular world.[15] Let us consider macromolecular chains composed of L- and D-monomers. For instance, we can think of nucleic acids, but it could be also a polypeptide or some other kind of unspecified polymeric macromolecule. To be pragmatic we will choose RNA, given its relevance in prebiotic scenarios such as the "RNA world."[16] The monomers would be ribonucleotides such as AMP, GMP, CMP and UMP, in this or any other suitable form, perhaps activated, but constructed of both enantiomers of ribose, or in other words racemic. The number of possible sequences of L- and D-links in the polymeric macromolecule is 2^n, where n is the number of nucleotides. All the sequences are chiral. As seen previously, for chains not exceeding a certain critical length—which depends on the initial amount of monomer, but can be assumed to be roughly 50—the whole spectrum of sequences is present, including the homochiral sequences. This has been called the chemical level of complexity, and the system is racemic at this stage. As the number of nucleotides per chain (n) rises, the statistical constraints are severe. The number of statistical sequences rises exponentially and can become astronomical, to the extent that only some specific sequences are represented, and most sequences do not exist at all. Thus, for chains of 150 nucleotides or more, almost every sequence is unique, in so much as the whole universe is too small to accommodate all the possible sequences. As seen previously for synthetic polymers, the chance of finding in one sample the two enantiomers of a given chiral chain is in effect zero. This level of evolutionary macromolecular complexity has been called the biological level. At the current stage of evolution in the biosphere, we only observe homochiral sequences, which are merely one particular kind among all possible sequences. Polymer chains of RNA and DNA are known to be matrices for self-replication, making complementary replicas. In Figure 2.4a we represent a fragment of a chain of RNA ($\cdots 3'$-GAGC-$5'\cdots$) that is serving as template for a complementary strand ($\cdots 5'$-CUC-$3'$) which is growing at the $3'$ end. Guanosine monophosphate (GMP) forms hydrogen bonds ($\cdots$) with the complementary base on the main strand (i.e., with cytosine, C) and a covalent phosphate link is eventually formed. Both strands have a helical structure and comprise nucleotides of the same chirality sign, i.e., they are formed of D-ribose. The polymerization can continue in this

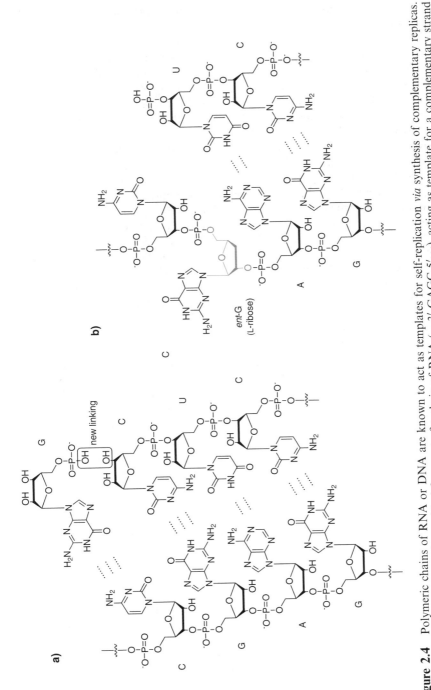

Figure 2.4 Polymeric chains of RNA or DNA are known to act as templates for self-replication *via* synthesis of complementary replicas. (a) Represents a homochiral fragment of a chain of RNA (···3'-GAGC-5'···), acting as template for a complementary strand (···5'-CUC-3') which is growing at the 3' end. If a chiral defect occurs, such as in (b) with the incorporation of L-ribose in the guanosine residue (*ent*-G), the matrix properties and replication abilities are disrupted.

way and a complementary RNA chain is eventually produced. Further replication of the complementary chain yields the original chain sequence, so actual replication has taken place every two generations. In a different RNA fragment, Figure 2.4b, the main strand has a chiral defect, in which one D-ribose has been substituted by L-ribose in the guanosine monophosphate nucleotide (*ent*-G), giving rise to the sequence (···3′-GA*ent*-GC-5′···). As a result of the chiral defect, the guanine base attached to L-ribose is at a large angle from its normal position in a homochiral chain, such as that in Figure 2.4(a). An estimate made by modeling polythymidine sets this angle near 100 °C.[15] The situation is represented in Figure 2.4b. The nitrogenous base of a chiral defect can not direct the coupling with the incoming complementary nucleotide and the polymerization is stopped, at least in the vicinity of the chiral defect. The important inference from this is that chains with a chiral defect completely lose their matrix properties and replication ability. These same conclusions are experimentally sustained by *in vitro* experiments of matrix oligomerization of nucleotides.[17] On the other hand, selection of sequences is only possible when the precision of the replication exceeds a certain threshold, otherwise errors catastrophe occurs.[18] The mean number of errors in a copy is $n \times q$, in accordance with the properties of the binomial distribution, where q is the probability of error at a single stage of chain enlargement. Therefore, the statistical pressure restrains the chain length at which mirror-symmetry breaking might have occurred during progressive evolution of structural and functional complexity. At a stage of evolution intermediate between the chemical and biological level, *i.e.*, for chain lengths between $50 < n < 150$, known as the prebiotic level of evolutionary complexity, the mirror-symmetry breaking might have appeared. Under these premises, the choice of L- or D-homochirality is intrinsic to the evolutionary process from the chemical to the prebiotic to the biological level of complexity, and is therefore spontaneous and stochastic in sign. It could be said that life on Earth makes and uses only one enantiomeric form, probably because anything else would excessively complicate the key biochemical processes.[19]

There is some experimental evidence, at least at the chemical level of complexity, for this proposal. Synthetic pyranosyl oligonucleotides are capable of self-assembling in a matrix-directed oligomerization.[20] As a result, homochiral D-libraries and L-libraries of higher oligomers are predominantly assembled from a mixture of all possible diastereomers of the initial smaller oligomeric units, without the participation of an enzyme. The theory is sound, but it would benefit from further experimental study involving analysis of increasingly complex mixtures. Spectacular advances in structural and molecular biology have added support to the RNA world hypothesis.[21] It may be also observed that, in spite of the indisputable successes of the RNA world,[16] there is no general consensus on the viability of this model and the chemical prebiotic reactions involved.[22] Difficulties in achieving a prebiotically plausible synthesis of RNA[23] have led many to ponder the possibility of a secondary post-biotic role of RNA,[24] and challenge the chemists to provide a means for a chemical explanation of the prebiotic generation of RNA in the early Earth.

Regarding different prebiotic macromolecular protagonists also capable of undertaking this evolutionary pathway, the polypeptides come to mind. Traditionally RNA has been regarded as having preceded proteins, partly due to the difficulty in composing scenarios in which proteins could replicate in the absence of nucleic acids. There are now examples showing that this is not necessarily the case, and self-replicant peptides are also described in laboratory *in vitro* experiments (Section 5.2.5, *Self-replication*).[25]

These evolutionary scenarios which implicitly embody both mirror-symmetry breaking and homochirality attributes are serious candidates in the ongoing debate reviewed in this book. Their success is intimately linked to a deeper understanding of prebiotic mechanisms, which are at best poorly understood and are currently difficult to verify. Unfortunately, as far as is known there is no remnant or evidence of precellular life anywhere on Earth, to the point that its existence could be regarded as entirely conjectural. If it did exist, the discovery of such remains from Archean fossils would provide precious, perhaps definitive, information.

In conclusion, both types of chance theories of the origin of biomolecular homochirality, the amplification of stochastic imbalances and polymerizations above a critical chain length, place a fundamental importance on the amplification of chirality (Chapter 5, *Mechanisms of Amplification*) which is carried out by an autocatalytic process (Section 5.2, *Autocatalysis*). This mechanism of amplification must be robust, since it starts working on what have been named cryptochiral systems,[9,26] or systems in which chirality is below detection limits (*e.g.*, one self-replicant polymeric molecule out of a whole mixture of diastereomers, or the standard deviation from purely racemic mixtures).

2.3 Deterministic Theories

In deterministic theories there is an initial chiral influence, regardless of its strength, which creates an enantiomeric imbalance in the originally achiral primeval matter. This chiral influence is responsible for the ultimate sign imprinted on the biomolecules. This disruption of symmetry is customarily named, as in the former case, the mirror-symmetry breaking step, and is the critical step towards homochirality. Once some kind of imbalance has been created, classical mechanisms, such as different solubility[27] or sublimation rates,[28] of racemates (Section 5.6, *Amplification of Scalemic Compounds: Eutectic Mixtures*) in addition to new proposed ones, such as the still theoretical Salam phase transition mechanism (Section 5.5, *The Salam Phase Transition*)[29] are known to operate, or might operate in further chiral enrichment and perhaps ultimately towards homochirality. This second process, also secondary in relevance, is the chiral amplification (Chapter 5, *Mechanisms of Amplification*). Although conceptually not the key step, this is also important and has received much attention in recent times, and deserves adequate coverage throughout this book. Once a set of molecules with sufficient

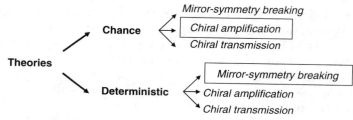

Scheme 2.2 Chance and deterministic theories of biomolecular homochirality, although necessarily evolving through the same sequential stages, mirror-symmetry breaking followed by chiral amplification, to which chiral transmission can be also added, differ in the importance of each step. Mirror-symmetry breaking is the key process in deterministic mechanisms, whereas this is not the case in chance mechanisms, in which an efficient mechanism of amplification of chirality, indeed amplification of cryptochirality, is the crucial process.

enantiomeric enrichment is produced, transference of their chirality to other molecules, with more or less efficiency, is the customary process. This chiral transmission, which is seen to have an enormous field of application in asymmetric synthesis, is in fact a rather trivial step. This third step is also third in relevance, and is not covered in the present text (Scheme 2.2).

Regardless of taxonomy and the need to focus on facts, we shall report in the following sections of the book some of the most relevant experimental data on mirror-symmetry breaking events, along with the deterministic theories formulated in their connection and a description of the chiral amplification mechanisms which have become better established. All of these are necessary to achieve a broad picture of one of the most highly captivating enigmas in Nature.

References

1. (a) J. Podlech, *Cell. Mol. Life Sci.*, 2001, **58**, 44–60; (b) U. Meierhenrich, W. H-P. Thiemann and H. Rosenbauer, *Chirality*, 1999, **11**, 575–582.
2. W. A. Bonner, P. R. Kavasmaneck, F. S. Martin and J. J. Flores, *Orig. Life*, 1975, **6**, 367–376.
3. (a) S. F. Mason, *Nature*, 1984, **311**, 19–23; (b) A. J. Macdermott, *Orig. Life Evol. Biosph.*, 1995, **25**, 191–199.
4. K. Evgenii and T. Wolfram, *Orig. Life Evol. Biosph.*, 2000, **30**, 431–434.
5. W. A. Bonner, *Top. Stereochem.*, 1988, **18**, 1–96.
6. Y. Nagata, in *Advances in Biochirality*, ed. G. Palyi, C. Zucchi and L. Caglioti, Elsevier Science BV, Amsterdam, 1999, 271–283.

7. T. Yoshimura and N. Esaki, *J. Biosci. Bioeng.*, 2003, **96**, 103–109.
8. (a) R. Plasson, D. K. Kondepudi, H. Bersini, A. Commeyras and K. Asakura, *Chirality*, 2007, **19**, 589–600; (b) D. G. Blackmond, *Prod. Natl. Acad. Sci. USA*, 2004, **101**, 5732–5736.
9. J. S. Siegel, *Chirality*, 1998, **10**, 24–27.
10. E. W. Weisstein, *Binomial Distribution*, from *MathWorld–A Wolfram Web Resource*; http://mathworld.wolfram.com/BinomialDistribution.html.
11. E. W. Weisstein, *Normal Distribution*, from *MathWorld–A Wolfram Web Resource*; http://mathworld.wolfram.com/NormalDistribution.html.
12. (a) G. Natta, *J. Polym. Sci.*, 1955, **16**, 143–154; (b) G. Natta, *Angew. Chem.*, 1956, **68**, 393–403.
13. M. M. Green and B. A. Garetz, *Tetrahedron Lett.*, 1984, **25**, 2831–2834.
14. *Aldrich Catalog Handbook of Fine Chemicals*, 2007–2008, Sigma–Aldrich Co., St. Louis, Missouri, USA.
15. V. Avetisov and V. Goldanskii, *Prod. Natl. Acad. Sci. USA*, 1996, **93**, 11435–11442.
16. (a) W. Gilbert, *Nature*, 1986, **319**, 618–618; (b) L. E. Orgel, *Crit. Rev. Biochem. Mol. Biol.*, 2004, **39**, 99–123; (c) T. R. Cech, *Int. Rev. Cytol.*, 1985, **93**, 3–22; (d) G. F. Joyce, *Nature*, 1989, **338**, 217–224; (e) A. J. Zaug and T. R. Cech, *Cell*, 1980, **19**, 331–338; (e) F. H. C. Crick, *J. Mol. Biol.*, 1968, **38**, 367–379; (f) L. E. Orgel, *J. Mol. Biol.*, 1968, **38**, 381–393.
17. G. F. Joyce, G. M. Visser, C. A. van Boeckel, J. H. van Boom, L. E. Orgel and J. van Westrenen, *Nature*, 1984, **310**, 602–604.
18. M. Eigen, J. McCaskill and P. Schuster, *J. Phys. Chem.*, 1988, **92**, 6881–6891.
19. P. Cintas, *Angew. Chem., Int. Ed.*, 2007, **46**, 9143–9144.
20. M. Bolli, R. Micura and A. Eschenmoser, *Chem. Biol.*, 1997, **4**, 309–320.
21. (a) G. F. Joyce, *Angew. Chem., Int. Ed.*, 2007, **46**, 6420–6436; (b) P. A. Monnard, *Orig. Life Evol. Biosph.*, 2007, **37**, 387–390.
22. (a) R. Shapiro, *Orig. Life*, 1984, **14**, 565–570; (b) R. Shapiro, *Orig. Life*, 1988, **18**, 71–85; (c) R. Larralde, M. P. Robertson and S. L. Miller, *Proc. Natl. Acad. Sci. USA*, 1995, **92**, 8158–8160; (d) R. Shapiro, *Orig. Life Evol. Biosph.*, 1995, **25**, 83–98.
23. (a) D. De Lucrezia, F. Anella and C. Chiarabelli, *Orig. Life Evol. Biosph.*, 2007, **37**, 379–385; (b) S. D. Copley, E. Smith and H. J. Morowitz, *Bioorg. Chem.*, 2007, **35**, 430–443.
24. C. Anastasi, F. F. Buchet, M. A. Crowe, A. l. Parkes, M. W. Powner, J. M. Smith and J. D. Sutherland, *Chem. Biodivers.*, 2007, **4**, 721–739.
25. A. Saghatelian, Y. Yokobayashi, K. Soltani and M. R. Ghadiri, *Nature*, 2001, **409**, 797–801.
26. K. Mislow and P. Bickart, *Israel J. Chem.*, 1976–1977, **15**, 1–6.
27. (a) Y. Hayashi, M. Matsuzawa, J. Yamaguchi, S. Yonehara, Y. Matsumoto, M. Shoji, D. Hashizume and H. Koshino, *Angew. Chem., Int. Ed.*, 2006, **45**,

4593–4597; (b) M. Klussmann, A. J. P. White, A. Armstrong and D. G. Blackmond, *Angew. Chem., Int. Ed.*, 2006, **45**, 7985–7989.
28. (a) P. Yang, R. Xu, S. C. Nanita and R. G. Cooks, *J. Am. Chem. Soc.*, 2006, **128**, 17074–17086; (b) S. P. Fletcher, R. B. C. Jagt and B. L. Feringa, *Chem. Comm.*, 2007, 2578–2580.
29. A. Salam, *Phys. Lett. B.*, 1992, **288**, 153–160.

CHAPTER 3
The Concept of Chirality

3.1 Introduction

In 1848 Pasteur deduced that his dissymmetry, or the *"property by which objects differ only as an image in a mirror differs from the object that produces it"*, had a molecular basis, and proposed that the observed biomolecular homochirality had its origins in the dissymmetric forces of Nature (Chapter 1, *Introduction and Historical Background*).[1] Following the discovery of the Faraday effect, or magnetically-induced optical rotation, in 1846,[2] Pasteur grew normally symmetrical crystals in a magnetic field with the object of inducing chirality in their crystal form. In addition, since he regarded spin as a dissymmetric force, he attempted to induce chirality by means of rotating devices. Evoking the movement of the solar system, he attempted to induce optical activity in synthetic products by carrying out chemical reactions in a centrifuge, and tried to modify the optical activity of natural products by rotating the plants producing them in a clockwise direction. As we are about to see, neither simple rotation nor magnetic fields are chiral influences, and neither the Earth's rotation nor its magnetic field were the chiral forces in Nature that Pasteur was seeking.

Among scientists, chemists are some of the ones most familiar with the concept of chirality. The term chiral is derived from the Greek *kheir*, meaning "hand" and, evoking Pasteur's dissymmetry, the expression was officially adopted by Lord Kelvin in 1904. In his Baltimore Lectures on Molecular Dynamics and the Wave Theory of Light, he stated, *"I call any geometrical figure, or group of points, chiral, and say it has chirality, if its image in a plane mirror, ideally realized, cannot be brought to coincide with itself"*.[3] Equivalent to this, but expressed in group theory language, the criterion for an object to be chiral is that it must not contain improper symmetry elements. This means the absence of improper rotation axes, S_n. Usually $S_1 = \sigma$ (plane of reflection) and $S_2 = i$ (centre of inversion) are the most commonly encountered attributes of symmetry in achiral molecules, although these are not the only possibilities.

The Origin of Chirality in the Molecules of Life
Albert Guijarro and Miguel Yus
© Albert Guijarro and Miguel Yus, 2009
Published by the Royal Society of Chemistry, www.rsc.org

The object, *e.g.*, a molecule, consequently must belong to one of the chiral point groups C_n, D_n, O, T or I.

In general we use the term chiral to refer to molecules and their two- or three-dimensional representations. However, chirality does not only apply to stationary objects such as these, but also to time-dependent physical entities represented by vectors and vector fields, either static or changing, some of them originating in translational or rotational movement. Under this premise, the traditional concept of chirality in chemistry was re-examined and broadened in 1986 by L D Barron. He proposed the terms "true chirality" and "false chirality" in a more extended definition of chirality.[4] According to this, two conditions must be fulfilled for a system to be truly chiral:

> A system exhibits true chirality if there exists a second system, known as an enantiomer, which
>
> > (a) is obtained by space inversion of the former and they are not superimposable, and
> > (b) the two systems are not interconverted by time reversal.

Within this definition, the absence of the second requirement leads to the condition known as "false chirality". We shall return to these concepts and their applications in Sections 3.3.1, 3.3.2 and 3.3.3, *Experiments in the Absence of Chirality* and *Experiments under the Influence of False* and *True Chirality*, respectively.

3.2 The Basic Symmetry Operations

A proper understanding of molecular chirality and its interaction with physical fields requires more comprehensive arguments than those of a simply geometrical nature, derived from point group symmetry and considered under the classical definition of chirality. In the following sections we shall explain the various symmetry effects in physical systems, either static or motion-related, including space inversion and time reversal, and the third symmetry operator, charge conjugation. It should be pointed out that the symmetry properties associated with these operators, discussed here from the standpoint of classical physics, can be extended to include those encountered in a quantum-mechanical treatment of the system, with the corresponding quantum-mechanical operators and expected values and associated symmetries.

3.2.1 Space Inversion: The Parity Operator, $\hat{P}$

The operation space inversion consists in inverting the system through the origin of the space-fixed axis; every point is projected to the opposite octant of Cartesian space. All the constituent particles defined by their position vectors $\vec{r}$ are moved to $-\vec{r}$. This operation converts an object to its enantiomer. This is equivalent to a reflection in a mirror containing the coordinate origin, followed

The Concept of Chirality

by rotation through 180° about an axis perpendicular to the plane of reflection. Since rotation invariance is a direct consequence of conservation of angular momentum, parity symmetry simply compares an object or a process with its mirror image, a concept familiar to chemists. The same final deductions can be drawn both from a space inversion or a mirror reflection treatment in the following explanations, the former being preferred for its mathematical simplicity. Most physical laws, particularly those of electromagnetism, are unchanged by space inversion. In other words, the equations representing the physical laws are unchanged if the space coordinates (x, y, z) are replaced throughout by $(-x, -y, -z)$, and the corresponding physical processes are said to conserve parity.

Vectors, in addition to scalars and tensors, are classified in two types according to their behavior under symmetry operations. Those vectors which change its sign under $\hat{P}$ are known as polar or true vectors, e.g., position vector $\vec{r}$ (Figure 3.1a) or $\vec{v}, \vec{E}, \vec{G}, \vec{p}$ (speed, electric field, gravitational field, linear momentum). A vector which does not change sign in the same operation is described as an axial or pseudo vector, e.g., angular momentum $\vec{L}$ (Figure 3.1b) or $\vec{B}$ (magnetic field). It can be deduced from elementary algebra that the axial vectors, originated by the vector product of two polar vectors (e.g., $\vec{L} = \vec{r} \times \vec{p}$), do not have their sign changed under the space inversion of their constituent polar vectors ($\hat{P}\vec{L} = -\vec{r} \times -\vec{p} = \vec{L}$). The parity operator leaves axial vectors unchanged, and this can be deduced from the unchanged sense of rotation after space inversion illustrated in Figure 3.1b.

3.2.2 The Time Reversion: The Time Reversal Operator, $\hat{T}$

On the other hand, the operation time reversal $\hat{T}$ reverses the absolute direction of time flow. It reverses the motion of all the constituent particles in the system, in other words it replaces the time coordinate t by $-t$ in all equations describing physical laws. Rewinding a movie of the system may give us a clearer idea of the

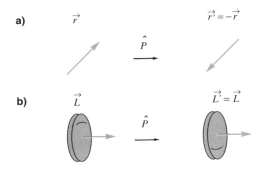

a) Polar vectors: $\vec{r}, \vec{v}, \vec{E}, \vec{G}, \vec{p}$ b) Axial vectors: $\vec{L}, \vec{B}$

Figure 3.1 (a) The effect of the parity operator $\hat{P}$ on the sign of the position vector $\vec{r}$ is to change its sign; (b) This has no effect on the angular momentum vector $\vec{L}$.

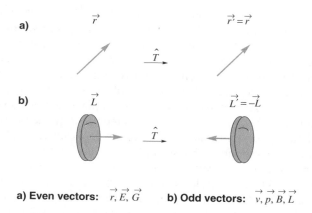

Figure 3.2 (a) The effect of the time operator $\hat{T}$ on the sign of the position vector $\vec{r}$ is to leave it unchanged; (b) It changes the sign of the angular momentum vector $\vec{L}$.

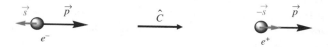

Figure 3.3 The effect of the charge conjugation operator $\hat{C}$ on a moving electron is to change the sign of the electric charge and magnetic moment of spin $\vec{s}$. The particle becomes an antiparticle.

effect of $\hat{T}$. Those processes which are unchanged by $\hat{T}$ are said to conserve or be invariant under time reversal. A vector with a sign that is not changed by $\hat{T}$ is referred to as time-even, e.g., the position vector $\vec{r}$ (Figure 3.2a), whereas a vector which changes its sign under operation $\hat{T}$ is said to be time-odd, for example angular momentum $\vec{L}$, as in Figure 3.2b. Quantities which do not depend explicitly on time, such as electrostatic or gravitational fields $(\vec{E}, \vec{G})$, are invariant under time reversal. Those that depend on the first derivative with respect to time, such as $\vec{v} = d\vec{r}/dt$, are time-odd in respect of time reversal: $\hat{T}\vec{v} = d\vec{r}/d(-t) = -\vec{v}$. It follows that those quantities proportional to the second derivative with respect to time, such as acceleration and force, $\vec{F} = m\vec{a} = md^2\vec{r}/dt^2$, are time-even.

3.2.3 The Charge Conjugation: The Charge Conjugation Operator, $\hat{C}$

This transformation is associated with the absolute sign of electric charge. $\hat{C}$ interconverts elementary particles (*e.g.*, electrons) into antiparticles (antielectrons or positrons) by changing their charge sign. Other physical properties connected to the electric charge (such as magnetic fields) are also affected in an equivalent manner (Figure 3.3). This might seem less relevant to chemists, who

The Concept of Chirality

are invariably concerned with matter, but it is important in high energy physics, where antimatter, or at least antiparticles, are present. We will return to this point later when we place emphasis on the meaning of true enantiomers.

3.3 The Failure of the Experiments of Pasteur, Curie and Others

3.3.1 Absence of Chiral Field

Considering these factors together, we are now in a position to see why Pasteur's experiments were doomed to fail. His chiral forces of Nature were the Earth's rotation and magnetic field. In vectorial notation, the Earth's rotation can be represented by its angular momentum vector $\vec{L}$, and $\vec{B}$ is the Earth's magnetic field; the magnitude of each depends on the rotational movement either of mass or charge, the sense of rotation being represented by a curved arrow (Figure 3.4a). If we apply the space inversion operator $\hat{P}$ to these axial vectors, $\vec{B}$ (or $\vec{L}$) remains unchanged. They have therefore no chiral influence and can not induce chirality under experimental conditions. In the case of polar vectors the lack of chirality of the field is evident *per se*. Following a similar argument, in an electrical $\vec{E}$ or gravitational field $\vec{G}$, while these are inverted by $\hat{P}$, the inverted field does not represent a chiral influence since it can be superimposed on the original vector simply by rotation through 180°, Figure 3.4b.

3.3.2 Experiments Under False Chiral Influence

In the previous case, none of the vector fields on their own are chiral, as Pasteur's contemporary Faraday observed. However, in 1894 Pierre Curie re-examined Pasteur's experiments on chiral forces in Nature and showed that there were two enantiomorphous combinations of collinear electric and magnetic fields, namely the parallel and the antiparallel alignment of fields, and that these could therefore be considered as chiral fields.[5] As represented in Figure 3.5a, a parallel set of electric and magnetic field, $\vec{E}$ and $\vec{B}$, is space-inverted by parity $\hat{P}$ to afford an antiparallel combination $-\vec{E}$ and $\vec{B}$, which is not superimposable on the former set, a *sine qua non* condition for chirality. However, both systems are examples of

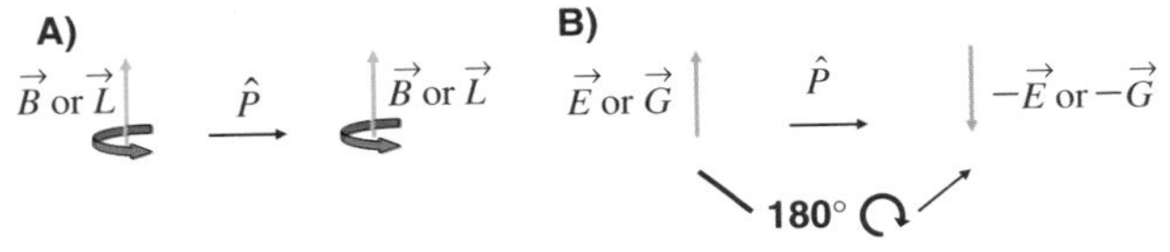

Figure 3.4 (a) Space inversion ($\hat{P}$) has no effect on magnetic field $\vec{B}$ nor on angular moment (rotation) $\vec{L}$, which are not chiral influences; (b) Evidently, neither the electric or gravitation fields are chiral, but following the same argument, the space-inverted field is superimposable on the original field by 180° rotation.

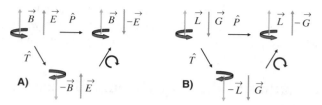

Figure 3.5 (a) The combination of a parallel electric and magnetic field ($\vec{E}\,\vec{B}$) gives under space inversion ($\hat{P}$) a different set of fields with antiparallel orientation ($-\vec{E}\,\vec{B}$), which is not superimposable on the former set; they can still be interconverted by time reversion ($\hat{T}$) and a 180° rotation (curved arrow), and they are therefore said to exhibit false chirality; (b) A combination of a rotation and gravity, $\vec{L}$ and $\vec{G}$, is formally analogous to the above system, displaying only false chirality.

false chirality, since they do not fulfill the second demand pointed out by Barron: non-interconvertibility by time reversion. Both sets of fields can be interconverted by the time reversal operator $\hat{T}$ (which in Figure 3.5b yields $\vec{E}$ and $-\vec{B}$) followed by 180° rotation (which affords the parity inverted system, $-\vec{E}$ and $\vec{B}$), and therefore means that they do not constitute a true chiral system. Exactly the same argument holds for $\vec{L}$ and $\vec{G}$ (angular momentum and gravity, represented in Figure 3.5b), for example in the hypothetical experiment of a frictionless spinning fluid in a gravitational field, in which case only false chirality is expected. An actual experiment under real conditions exploiting this combination of fields is reported in Section 4.4, *Fluid Dynamics: Vortex Motion*.

3.3.3 Experiments Under True Chiral Influence

The search for a true chiral field, and experimental evidence of its effects, yielded fruits only recently, even though it had been suggested many years ago.[6] The experiment is described as magnetochiral dichroism,[7] in which an unpolarized light beam (laser) refracts,[8] or photolyzes asymmetrically,[9] in its interaction with a racemic mixture of Eu((±)tfc)$_3$ [europium(III) *tris*-(3-trifluoroacetyl-(±)-camphorate)] or [Cr(ox)$_3$]$^{3-}$ [chromium(III) *tris*-oxalate], in the first and second case, respectively, when these complexes are placed under a powerful static and collinear magnetic field. Opposite chiral effects were observed when the laser beam and magnetic field were *parallel or antiparallel* (for further details, see Section 4.3.2, *The Magnetochiral Effect*). Again, the experiment can be regarded as the combined effect of a time-odd polar vector, $\vec{k}$ (propagation of light, *i.e.*, analogous to linear momentum) and a time-odd axial vector, $\vec{B}$ (magnetic field). The resulting field ($\vec{k}\,\vec{B}$, Figure 3.6) is not superimposable on the field obtained after space inversion ($\hat{P}$), *i.e.*, it is chiral, and can not be interconverted by time reversal (which gives $-\vec{k}\,-\vec{B}$) plus any appropriate rotation.

A similar treatment can be applied to elementary particles moving ($\vec{p}$) and spinning ($\vec{s}$). Let us consider the electron ($s = 1/2$, $m_s = \pm 1/2$), with its spin angular momentum vector projection either parallel or antiparallel to the

The Concept of Chirality

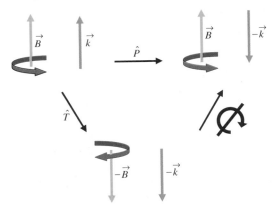

Figure 3.6 The field originated combining a parallel light beam and a magnetic field, $\vec{k}\vec{B}$, gives under space inversion ($\hat{P}$) a different array of fields with antiparallel orientation, $-\vec{k}\vec{B}$, which are enantiomorphous. These are truly chiral fields since can not be interconverted by time reversal ($\hat{T}$) plus any appropriate rotation (slashed curved arrow).

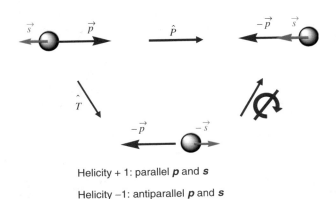

Helicity + 1: parallel **p** and **s**

Helicity −1: antiparallel **p** and **s**

Figure 3.7 Elementary particles such as the electron have two non-superimposable quantum states with either parallel (helicity +1) or antiparallel (helicity −1) arrays of linear and angular momentum. They are truly chiral systems, since these states can not be interconverted by time reversal plus any appropriate rotation.

propagation direction as its two possible quantum states (Figure 3.7). Space inversion (parity operator $\hat{P}$) interconverts both systems. Also, they are not superimposable and they are not interconverted by time reversal plus any appropriate rotation. The system displays true chirality, with opposite spin projections corresponding to opposite handedness. The term helicity is reserved to denote the handedness of moving and spinning particles. Formally endowed with chirality, helicity −1 corresponds to the antiparallel array of momenta and helicity +1 to the parallel array.

3.4 Expected Effect of True and False Chirality on Molecules

What is then the effect of a false chiral influence on the energy—and therefore the properties—of enantiomeric molecules? And what is the effect of a true chiral influence? A key difference between true and false chiral influences becomes apparent when we consider two separate situations, namely states which have reached thermodynamic equilibrium and are therefore time invariant, and on the other hand processes under kinetic control. Only under the effect of a true chiral field will the energy of a chiral molecule be different from that of its enantiomer. In other words, enantiomers remain strictly degenerate in the presence of a false chiral influence. However, it has been pointed out that in theory a falsely chiral influence can modify asymmetrically the activation energy barriers which lead to the synthesis of enantiomers in processes under kinetic control. Starting from non-optically active reagents, the result of this would be an asymmetric synthesis, provided that thermodynamic equilibrium was not reached.[10] This conclusion is significant since it implies breakdown of microscopic reversibility, but only under the effect of a false chiral field. An example of this is shown in Figure 3.8 within the context of an electrocyclic reaction. The conrotatory interconversion of a butadiene and two enantiomeric chiral cyclobutenes is a good theoretical example used by Barron to illustrate this effect.[11]

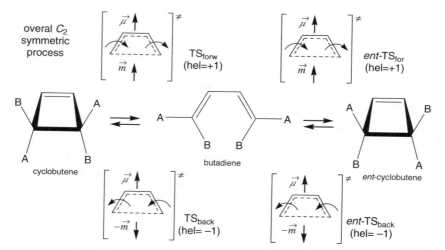

Figure 3.8 During the [2 + 2] electrocyclic ring-closing of a substituted butadiene to give enantiomeric cyclobutenes, a transient array of either parallel ($\vec{\mu} + \vec{m}$) or antiparallel ($\vec{\mu} - \vec{m}$) electric and magnetic momenta appear, depending on the sense of rotation of the process. While the products and reagents remain strictly degenerate, even in the presence of a false chiral influence, the transition state degeneracy is lost by the effect of a false chiral field (e.g., a collinear electric and magnetic field), opening the possibility of an asymmetric synthesis under kinetic control.

The Concept of Chirality 29

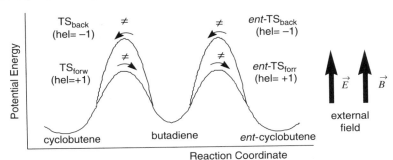

Figure 3.9 Under the effect of a collinear electric and magnetic field, the potential energy profiles for the experiment in Figure 3.8 are altered; activation energies are lowered for the forward reactions and increased for the backward reactions (curved arrows). Parallel ($\vec{\mu} + \vec{m}$, hel = +1) and antiparallel ($\vec{\mu} - \vec{m}$, hel = −1) arrangement of transient momenta interact differently, with the external field splitting the degeneracy in the corresponding transition states. This is formally a breakdown of the principle of microreversibility, since forward and backward reactions have different barriers. Theoretically, this phenomenon could find application in asymmetric synthesis under kinetic control. Note that products and reagents remain strictly degenerate in the presence of any falsely chiral field, so thermodynamic equilibration is not altered.

The entire process is a C_2-invariant transformation. Along the reaction coordinate pathway, symmetry arguments show that, as electric density begins to accumulate between the ends of the diene through the transition state, an electric dipole moment $\vec{\mu}$ develops parallel to the C_2 axis. Simultaneously, the rotational motion of the groups A and B generates transient magnetic moments $\vec{m}$, also parallel to the local rotation axis, with equal magnitude and opposite sign depending on the direction of the rotation. From Figure 3.8 it can be seen that the forward and backward reactions involve parallel and antiparallel arrangements of transient electric and magnetic moments, respectively, i.e., $\vec{\mu} + \vec{m}$ for the reactions cyclobutene → butadiene → *ent*-cyclobutene and $\vec{\mu} - \vec{m}$ for the backward reactions, cyclobutene ← butadiene ← *ent*-cyclobutene. The sense of rotation and its associated magnetic moment ($\vec{m}$ or $-\vec{m}$) are represented in the transition states (TS) along with their transient helicities. It is now clear that an external field consisting of a parallel ($\vec{E}\vec{B}$) combination of a magnetic plus an electric field, i.e., a false chiral influence, will interact differently to the antiparallel arrangement ($\vec{E}-\vec{B}$). These transient pairs of moments have different potential energy barriers for the forward and backward reactions, as represented in the energy profile of Figure 3.9. This will results in a bias of one enantiomer over the other, provided that the reaction is under kinetic control. Although theoretically feasible, this form of kinetic resolution of enantiomers still lacks experimental validation. So far, we have no experimental evidence for the breakdown of microscopic reversibility in chemical reactions induced by a false chiral influence.

References

1. L. Pasteur, *C. R. Acad. Sci. Paris*, 1848, **26**, 535–539.
2. M. Faraday, *Philos. Mag.*, 1846, **28**, 294–317.
3. Lord Kelvin (W. H. Thomson), *Baltimore Lectures*, C. J. Clay and Sons, London, 1904, p. 619.
4. (a) L. D. Barron, *Chem. Phys. Lett.*, 1986, **123**, 423–427; (b) L. D. Barron, *Chem. Soc. Rev.*, 1986, **15**, 189–223.
5. P. Curie, *J. Phys. Théor. Appl.*, 1894, **3**, 393–415.
6. (a) N. B. Baranova, B. Ya Zel'dovich, *Mol. Phys.*, 1979, **38**, 1085–1098; (b) G. Wagnière and A. Meier, *Experientia*, 1983, **39**, 1090–1091.
7. L. D. Barron and J. Vrbancich, *Mol. Phys.*, 1984, **51**, 715–730.
8. G. L. J. A. Rikken and E. Raupach, *Nature*, 1997, **390**, 493–494.
9. G. L. J. A. Rikken and E. Raupach, *Nature*, 2000, **405**, 932–935.
10. L. D. Barron, *Chem. Phys. Lett.*, 1987, **135**, 1–8.
11. L. D. Barron, *Chem. Phys. Lett.*, 1994, **221**, 311–316.

CHAPTER 4
Chiral Physical Forces

4.1 Background

Experimental evidence for enantiomer discrimination by physical forces or related effects discernible at the macroscopic level is relatively scarce. A chiral physical field, or in the loose sense of the word a chiral physical force, can provoke a chiral perturbation in a chemical system, either racemic or prochiral, at some point along the reaction coordinate that may bring about an asymmetric chemical imbalance, *i.e.*, $ee \neq 0$. This absolute asymmetric synthesis, *i.e.*, the formation of enantiomerically enriched products from achiral precursors without the intervention of chiral chemical reagents or catalysts, has only been carried out by a limited number of physical agencies. Indeed, it has so far been demonstrated in only three cases. In two of these, it appears as a consequence of the asymmetric interaction of light upon matter. In chronological sequence, these are natural circular dichroism and magnetochiral circular dichroism, respectively (Section 4.3, *Asymmetric Photolysis and Photosynthesis*). The third physical chiral field which has actually induced chiral selection at the molecular level with macroscopic effect arises from the dynamic of fluids (Section 4.4, *Fluid Dynamics: Vortex Motion*). This has taken place despite the intrinsic chiral nature of the weak force and its omnipresent character; any measurable asymmetric chemical unbalance generated by this fundamental interaction remains elusive. The nature of all these physical chiral forces is discussed in this chapter.

4.2 The Weak Interaction and the Parity Violation

4.2.1 The Fundamental Interactions: Gravitational, Weak Interaction, Electromagnetic and Strong Interaction

In order to appreciate the significance of Pasteur's original suggestions (Chapter 1, *Introduction and Historical Background*) we need to review some of the

---

The Origin of Chirality in the Molecules of Life
Albert Guijarro and Miguel Yus
© Albert Guijarro and Miguel Yus, 2009
Published by the Royal Society of Chemistry, www.rsc.org

fundamental concepts of physics. From a global point of view, all the phenomena which occur in Nature can be explained in terms, or as a result, of four fundamental interactions or forces: gravitational, weak interaction, electromagnetic and strong interaction. Comprehension of these forces and the mechanisms by which they interact with the particles of matter is the ultimate purpose of the Standard Model.[1] The Standard Model arises from the unification of two theories of particle physics into a single framework to describe all interactions of subatomic particles (except those due to gravity, for which no antiparticle or force carrier is yet known). The two components of the Standard Model are electroweak theory, which describes interactions *via* the electromagnetic and weak forces, and quantum chromodynamics, the theory of the strong nuclear force. Forces act on matter. Within this model, the two basic constituents of matter are leptons and quarks, particles of spin = 1/2 (fermions), each of which are believed to exist in six types, plus the corresponding antiparticle, each of them organized into three generations of increasing mass. Their given names—odd in some instances—are summarized in Figure 4.1. Ordinary atoms are made of particles from the first generation, the lightest ones. Higher generation particles tend to decay into lower generation particles. The second and third generation of particles are only detected in experiments of high-energy physics. Although the existence of three generations was deduced theoretically (see Chapter 9, *Intrinsic Asymmetry of the Universe: The Arrow of Space–Time and the Unequal Occurrence of Matter and Antimatter*) there is some debate about its fundamentality, being possibly regarded as excited states of the first generation.

Leptons are elementary particles which do not undergo the strong interaction, the electron, e^-, being the most representative lepton. Unlike leptons, quarks do not reveal themselves alone, due to their strong interaction. They are always found in pairs or triplets with other quarks or antiquarks, giving rise to mesons and barions, a plethora of subatomic particles collectively known as hadrons. Protons and neutrons are hadrons and possess internal structure; the proton can be described as being formed by three quarks, uud (up, up and down quarks), whereas the neutron is udd (up, down, down).

The elementary particles of matter interact with one another through the four distinct types of force mentioned above. The Standard Model proposes that each force is originated by exchange of a virtual elementary particle: a boson, or force carrier of spin = 1 (Figure 4.2). Thus, the electromagnetic force is mediated by the photon, the basic quantum of electromagnetic radiation. The strong force is mediated by gluons and gravity, and although not part of the Standard Model is thought to be mediated by gravitons. These are all bosons. The weak force, which is our central subject of discussion, is mediated by the W^+, W^- and Z° bosons.

The nature of the electromagnetic force between two charges, **q** and **q'**, is explained as the result of the exchange of a virtual photon, γ (Figure 4.3, *left*). Transformations between subatomic particles are mediated by weak forces. Collisions and disintegrations are examples of weak interactions. A neutron (**n**) interacts with a neutrino (v_e) to give rise to a proton (**p**) and an electron (**e**⁻), an example of weak interaction mediated by the W^+ boson (Figure 4.3, *centre*). The β-decay in radioisotopes is another example of weak interaction. A neutron (**n**)

Chiral Physical Forces

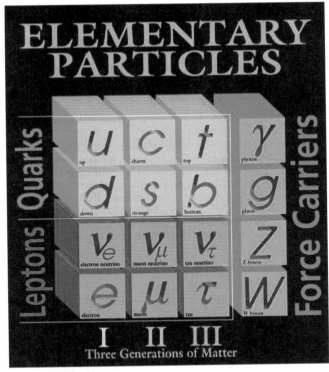

Figure 4.1 The Standard Model. The elemental particles which form the basis of matter are divided into two families, leptons and quarks, each organized in three generations. The most common leptons are the electron and its associated neutrino. Unlike leptons, quarks are not present alone but in groups, due to the strong interaction. Proton (uud) and neutron (udd) are the most common triads of quarks in matter. In addition to matter, another set of particles known as bosons are the carriers of the fundamental forces, photons for the electromagnetic forces, gluons for the strong nuclear forces and W and Z° for the weak interaction. (Picture from Fermilab: http://www.fnal.gov/pub/inquiring/matter/madeof/index.html.)

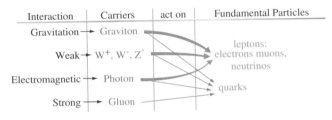

Figure 4.2 The four fundamental interactions (gravitational, weak interaction, electromagnetic and strong interaction) are considered to be the result of the exchange of virtual particles or force-carriers between the fundamental particles of matter: leptons and quarks.

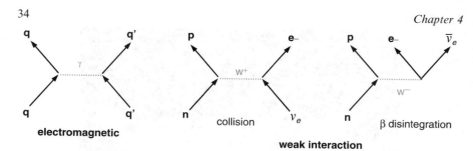

Figure 4.3 Interactions resulting from the exchange of force-carriers. Electromagnetic interaction between two charges, **q** and **q'**, as a result of a virtual photon exchange γ (*left*). Particle collisions and disintegrations are mediated by weak interactions. As the result of a collision between a neutron and a neutrino, a proton and an electron emerge, the reaction being mediated by the virtual carrier **W**$^+$ (*centre*). In the radioactive β-decay, a neutron disintegrates into a proton, an electron and an antineutrino in a charged current weak interaction mediated by **W**$^-$ (*right*).

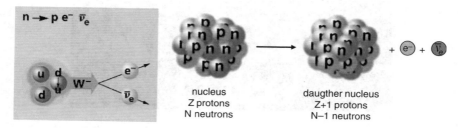

Figure 4.4 β-Decay: a neutron (udd quarks) decays to a proton (uud quarks), an electron and an antineutrino *via* virtual boson carrier **W**$^-$ (*left*). In a nucleus it originates a daughter nucleus with an atomic number increased by one unit (*right*).

spontaneously decays to a proton (**p**), an electron (e$^-$, *i.e.*, the actual β-radiation) and an antineutrino ($\bar{\nu}_e$) in a **W**$^-$-mediated charged current weak interaction (Figure 4.3, *right*).

4.2.2 The β-Decay

The β-decay is one of the processes of radioactive disintegration in which an unstable nucleus ejects an energetic electron and an antineutrino, while a neutron (udd) is transformed into a proton (uud), which remains in the product nucleus. This is an example of charged current weak interaction mediated by the weak boson **W**$^-$ (Figure 4.4, *left*). Thus, β-decay results in a daughter nucleus, the atomic number of which is one more than its parent but the mass number is unchanged (Figure 4.4, *right*).

Along with the β-radiation, which consists of a beam of high-energy left-handed electrons, an exotic particle called an *antineutrino* is released in this

Figure 4.5 Neutrinos are the simplest manifestation of chirality. Neutrinos are always left-handed (helicity −1), while antineutrinos are right-handed (helicity +1). Without mass or charge, and being unaffected by the strong force, they only participate in weak interactions.

process. Within the standard model, the neutrino has a zero mass at rest, zero charge and spin 1/2. It can be one of three different types or families as indicated above, one for each kind of charged lepton. Neutrinos and antineutrinos are the most elemental manifestation of chirality. The only difference between them is their helicity—or, in other words, their chirality. As represented in Figure 4.5, neutrinos only exist with helicity −1 (antiparallel linear momentum and spin) and antineutrinos with helicity +1 (parallel linear momentum and spin) (see Section 3.2, *The Basic Symmetry Operations*). The left-handed helicity of neutrinos was experimentally measured in 1958.[2]

4.2.3 Parity Violation

In 1927, E. P. Wigner established the principle of conservation of parity, by which *"all interactions in nature are invariant with respect to space inversion"*, i.e., parity is invariably conserved in all cases.[3] What this means is that all attractive or repulsive forces between charges or masses, and the dynamics, kinematics or any other manifestation associated with these forces, obey the same laws and are governed by the same rules on both sides of the mirror. This may seem self-evident now, and in fact it was probably then, given the state of knowledge of physics at the time. It is in fact a fundamental link between symmetry and the laws of Nature. With the discovery of the new interactions (weak and strong), things were about to change, however, and in 1956 T. D. Lee and C. N. Yang predicted that parity was theoretically not preserved in weak nuclear interactions.[4]

4.2.3.1 Experimental Confirmation of Parity Violation in Weak Interactions

Only one year after the Lee and Yang's theoretical prediction of the non-conservation of parity in weak interactions, C. S. Wu designed an experiment to illustrate parity violation experimentally.[5] The Wu experiment dealt with the β-decay of ^{60}Co in which, a neutron **n** decays into a proton **p**, an electron **e**$^-$ and electron antineutrino $\bar{\nu}_e$, as we saw in Section 4.2.1, producing ^{60}Ni (Figure 4.6). The nuclear spin magnetic moment of ^{60}Co nuclei $\vec{I}$ was aligned with an intense

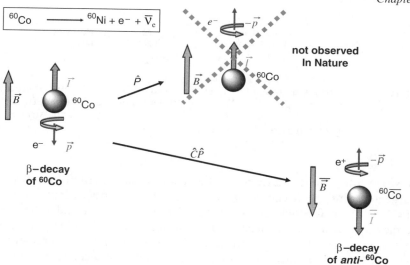

Figure 4.6 Wu's experiment: β-Decay of ^{60}Co nuclei is aligned with a magnetic field. The emission of β-radiation occurs through the south pole of the aligned nuclei (*left*). The space-inverted experiment ($\hat{P}$) can not be realized (*upper right*) and illustrates parity violation. Symmetry is recovered by applying space-inversion plus charge conjugation $\hat{C}\hat{P}$ to the original experiment, which affords anti-^{60}Co and reversed magnetic moment and field. This hypothetical experiment exploring the anti-world is equivalent to that observed in Nature.

external magnetic field $\vec{B}$ at a temperature close to absolute zero. The spatial distribution of the β-radiation was then measured. Left-handed electrons were found to be emitted preferentially in a direction antiparallel to the magnetic field, *i.e.*, through the "south pole" of the nuclei. The space-inverted experiment (Section 3.2.1, *Space Inversion: The Parity Operator, $\hat{P}$*), obtained after applying $\hat{P}$, is represented in Figure 4.6. $\vec{B}$, $\vec{I}$ and the curved arrow representing the electron spin (angular momentum) are axial vectors and do not change sign after applying $\hat{P}$, whereas the linear momentum of the electron $\vec{p}$ does so. In the space-inverted experiment electrons should be emitted parallel to the magnetic field, through the "north pole" of the nuclei. This is not observed in Nature and presents unequivocal evidence for parity violation. Interestingly, symmetry is recovered, invoking charge conjugation simultaneously with space inversion (Section 3.2.3, *The Charge Conjugation: The Charge Conjugation Operator, $\hat{C}$*). Under $\hat{C}\hat{P}$ operation, anti-^{60}Co emits right-handed positrons, and the external magnetic field and its spin magnetic moment are reversed: $\vec{\overline{B}}$, $\vec{\overline{I}}$. This hypothetical experiment belonging to the anti-world is an actual mirror image experiment, formally equivalent to the naturally observed β-decay. Anti-^{60}Co, like most antimatter, is not available for testing. The situation is however different in high-energy physics, where both particles and antiparticles are encountered routinely.

Chiral Physical Forces

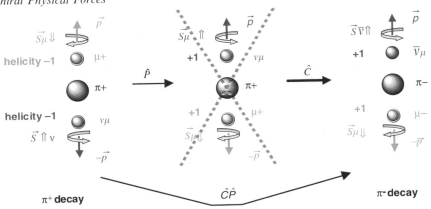

Figure 4.7 The meson π^+ decays naturally into a left-handed muon μ^+ and a left-handed neutrino v_μ (*left*). The space-inverted reaction ($\hat{P}$) is not found in Nature and evidences parity violation (*centre*, crossed in red). Instead, the space-inverted charge-conjugated process is permitted, corresponding to the natural decay of the meson π^- (*right*). This illustrates $\hat{C}\hat{P}$ conservation and the recovery of symmetry in processes mediated by weak interactions.

4.2.3.2 π^+ and π^- Decay and the Conservation of the Symmetry Operator, $\hat{C}\hat{P}$

The meson π^+ is a spinless subatomic particle found in cosmic rays, which decays into a muon (μ^+) and a muon neutrino (v_μ), as illustrated in Figure 4.7, *left*. Like all decay, it is a weak interaction (in this case, mediated by the force carrier, W^+). Conservation of linear momentum requires that the products of decay are emitted in opposite directions with equal and opposite momenta; meanwhile, conservation of angular momentum requires that they possess equal and opposite spin. The weak interaction governing all decay processes allows creation only of left-handed neutrinos, with negative helicity (*i.e.*, having their spin angular momentum ($\vec{s}$) and linear momentum ($-\vec{p}$) in opposite directions, as shown), and right-handed antineutrinos, with positive helicity. Consequently, the muon μ^+ must also have spin pointing back along its momentum. The space-inverted ($\hat{P}$) reaction (or mirror image) never occurs in nature, because the parity inversion performed by the mirror changes a left-handed neutrino to a right-handed state, which the weak interaction can not produce (Figure 4.7, *centre*). If, however, the $\hat{P}$ (parity) operator is combined with the $\hat{C}$ (charge conjugation) operator, the positive charges become negative and the neutrino changes to an antineutrino ($\bar{v}_\mu$), which is allowed (in fact required) to have positive helicity (*i.e.*, to be right-handed) and the reaction changed in this way is allowed just as much as the original. This is, in fact, the well-known meson π^- decay (Figure 4.7, *right*). This is only one of many experiments conducted since Wu's time which confirm parity violation $\hat{P}$, and validates the complementary symmetry operation of charge conjugation plus space-inversion, or $\hat{C}\hat{P}$ operation.

A central idea which emerges from these experiments is that if a system displays weak interaction it is not separately invariant under the parity operator $\hat{P}$ or charge conjugation, but it is invariant under the parity plus charge-conjugation operator $\hat{C}\hat{P}$, which restores the symmetry of the interaction. This is the principle of conservation of $\hat{C}\hat{P}$, which operates in *almost* all known processes in Nature.

4.2.3.3 The Principle of $\hat{C}\hat{P}\hat{T}$ Invariance

Examples in Nature can be quoted illustrating individual violation of each of the three symmetries, $\hat{P}$, $\hat{C}$ and $\hat{T}$.[6] It was at one time thought that $\hat{C}\hat{P}$ (parity transformation plus charge conjugation) would always leave a system invariant, but the notable example of the neutral meson K^0 (and more recently the B_0, see Chapter 9) has indicated a slight violation of $\hat{C}\hat{P}$ symmetry.[7] This small violation of $\hat{C}\hat{P}$ symmetry suggests some departure from $\hat{T}$ symmetry (Section 3.2.2, *The Time Reversion: The Time Reversal Operator, $\hat{T}$*). Unification of symmetry laws arises from the combination of all three symmetries, producing the $\hat{C}\hat{P}\hat{T}$ invariance principle.[8] The principle of $\hat{C}\hat{P}\hat{T}$ invariance requires that mirror image systems—with all objects inverted by parity inversion $\hat{P}$—and with all matter replaced by antimatter, corresponding to the operation of charge conjugation $\hat{C}$, should evolve in exactly the same way at any time. In the remote possibility that this is not achieved, every violation of the combined symmetry of two of its components, such as $\hat{C}\hat{P}$, must occur with a corresponding violation in the third component, $\hat{T}$. In other words, to preserve overall $\hat{C}\hat{P}\hat{T}$ symmetry, a $\hat{C}\hat{P}$-violating process implies full correspondence of, firstly, the original system in which that process is evolving with time and, secondly, the space-inverted, charge-conjugated system, but running backwards in time. Indeed, the violations in $\hat{T}$ symmetry are indistinguishably referred to as $\hat{C}\hat{P}$ violations and *vice versa*, and this can be extended to any pair of individual symmetries. In these $\hat{C}\hat{P}$-violating weak interactions the principle of microreversibility is not expected to hold, resulting in fundamental cosmological implications (see Chapter 9, *Intrinsic Asymmetry of the Universe: The Arrow of Space–Time and the Unequal Occurrence of Matter and Antimatter*). The invariance of $\hat{C}\hat{P}\hat{T}$ seems to stand on very firm ground. It represents a profound level of symmetry, consistent with all known experimental observations and obeyed by all the forces in Nature. Its endurance will be tested in successive generations of experiments in high energy physics.

4.2.4 Unification of Forces

The Standard Model considers the fundamental forces to be the result of particle exchange and includes in the same model the strong and weak interaction and the electromagnetic interaction. It is a step towards to the unification of all forces, which originated with Maxwell's study of electromagnetism. Later, the weak interaction was incorporated in what was called the "electroweak theory". Shortly afterwards, the strong interaction was also integrated. The electroweak theory is of enormous importance in all aspects of chirality and its primary origins.

4.2.4.1 The Electroweak Theory

In the 1970s physicists S. Glashow,[9] S. L. Weinberg[10] and A. Salam[11] concluded that the electromagnetic force and the force responsible for radioactive decay, the weak interaction, were themselves derivatives of a more unified field of influence, the electroweak force.[12] This unification of the quantum mechanical theories of weak and electromagnetic interactions had a direct implication—that there must be a new kind of weak interaction in addition to the well-known weak charged current (mediated by bosons W^+ and W^-), named weak neutral current, carried by the neutral boson, $Z°$. This neutral current generates parity-violating charge-preserving interactions between electrons ($e^- \cdots e^-$) and between electrons and quarks in the nucleus ($e^- \cdots$ up and down quarks) in stable matter, among many other interactions involving less common particles.

4.2.4.2 The Neutral $Z°$ Boson

The prediction of this particle from the electroweak theory gives rise to a whole new range of neutral current phenomena, interactions mediated by the $Z°$ boson in which the charge of the interacting particles is not altered, and which give rise to important implications for our discussion:

- The weak interactions are not limited to high-energy processes but would also take place in stable atoms (C, H, N, O, *etc.*), and therefore in biomolecules.[13]
- Two mirror-image enantiomeric molecules are energetically nonequivalent.[14] This lifting of the exact degeneracy of energy levels is due to parity violation associated with the weak neutral current.
- The true enantiomer of a molecule should be obtained by applying the $\hat{C}\hat{P}$ operator. It is a mirror-image molecule but constructed with antiparticles, belonging therefore to the domain of antimatter.[15]

The discovery of the $Z°$ boson (following those of W^+ and W^- bosons) took place at CERN in 1983 in a proton–antiproton collider,[16] in what must rank as one of the most important experiments in the history of science. The characteristics of these long-awaited particles were exactly as predicted by the electroweak theory. The discovery of the weak boson carriers confirmed the Standard Model with high accuracy and without revealing any flaws (Figure 4.8), a model that can be applied to the majority of interactions between particles.

Since that time, examples of weak neutral current in stable atoms have been illustrated experimentally in many ways,[17] and have so far validated the

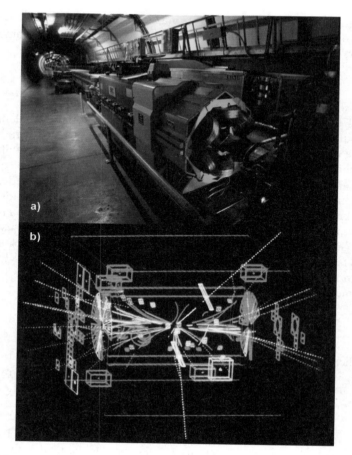

Figure 4.8 (a) Inside the LEP (Large Electron–Positron collider) tunnel at CERN. Within its 27 km circumference hundreds of events involving W± and Z° particles were studied, e.g., tracks left by the Z°; (b) Extremely precise measurements of W and Z° at CERN have confirmed the electroweak theory predictions beyond doubt. (Pictures from CERN: http://outreach.web.cern.ch/outreach/public/cern/PicturePacks/Picturepacks.html)

Standard Model theory.[18] In the form of optical activity, parity violation has been measured as tiny optical rotations with the correct sign and order of magnitude in vapors of metals (Bi,[19] Pb,[20] Tl,[21] Cs[22] and Pb[23]). The preponderance of heavy atoms is not accidental. Among the weak current interactions acting in stable atoms, those between electrons and quarks in the nucleus ($e^- \cdots$ quarks) provide the largest contribution. The development of a theory of parity violation in atoms has revealed that the strength of the weak neutral current increases with the third power of the atomic number, i.e., with Z^3.[24] Apart from optical activity, parity violation has also been experimentally observed in different sets of experiments, as parity-violating scattering at right- and left-handed high-energy electron beams by the quarks (nucleons) in hydrogen[25] and deuterium targets,[26] or broken mirror-symmetry during absorption of light (*e.g.*, laser excitation of the 6S–7S transition in Cs),[27] as well as in many other experiments in high-energy physics.

4.2.5 Quantification of the Parity Violation in Molecules: ΔE_{pv}

In molecules, effects involving weak neutral currents similar to those described for atoms are also present. However, since most organic molecules have a closed-shell configuration, the effect of the two opposite spins of the electrons in each molecular orbital causes the parity-violating interaction to vanish. This happens unless the molecule is chiral, in which case it is conceivable an imbalance in the energy of the electrons of opposite helicity (see Figure 3.7 and caption). This effect is attained through what is known as spin-orbit coupling, a weak magnetic interaction between the spin of the electrons and their orbital motion which slightly splits the otherwise identical (degenerate) energy of the two spin configurations. The resulting configuration of spin is of course reversed in each enantiomeric molecule. The electromagnetic field of forces created by the nucleus in the molecule is blind to the chirality of the interacting particles, but the weak force field is not. The introduction of the weak current interaction component $\hat{H}_{pv}$ to the molecular Hamiltonian—carrying the nuclear coordinates—provides a way to evaluate its effect on the energy by calculation. This operator can be formulated within the Standard Model framework from the expression describing the electroweak interaction energy between leptons and quarks.[28] After applying a number of approximations (including neglecting the $e^- \cdots e^-$ weak current interaction and taking the low energy semi-relativistic limit, among others) as well as neglecting the smaller nuclear spin-dependent term of the Hamiltonian expression, an indication of the parity-violating potential, $\hat{H}_{pv}$, is obtained, Equation (4.1).[29–33] The introduction of the spin-orbit coupling in the calculations reveals that the dependence of the weak current on the nuclear charge increases to a higher power, *ca.* Z^5.

$$\hat{H}_{pv} = \frac{G_F}{2m_e c \sqrt{2}} \sum_a^{nucl} Q_a \sum_i^{elect} [\vec{s}_i \cdot \vec{p}_i \delta(\vec{r}_{ia}) + \delta(\vec{r}_{ia}) \vec{s}_i \cdot \vec{p}_i] \qquad (4.1)$$

A closer look at this expression reveals certain important aspects of the interaction. The first term of Equation (4.1) is a constant; it comprises G_F, the Fermi weak coupling constant (a very small factor that determines the smallness of the interaction), m_e, the electron mass, and c, the speed of light. In the nuclear term, there is the electroweak charge, $Q_a = (1-4\sin^2\theta_W)Z_a - N_a$, where the Weinberg angle θ_W is another constant, and Z_a and N_a represent the proton and neutron number of nucleus, a. The electronic term is of special relevance to us. It includes the Dirac delta distribution, $\delta(\vec{r}_{ia})$, which confines the interaction of the electron i, of relative coordinates $\vec{r}_{ia}$ with respect to the nucleus a, to the point of the nuclear position (i.e., $\delta(\vec{r}_{ia})$ is zero everywhere except at the site of the nucleus) and, most importantly, the scalar product of the spin angular momentum operator $\vec{s}_i$ and the linear momentum operator $\vec{p}_i$ for the electron i: $\vec{s}_i \cdot \vec{p}_i$. As seen in Section 3.3.3, *Experiments under True Chiral Influence*, the spin angular momentum is a time-odd axial vector, whereas the linear momentum is a time-odd polar vector. The scalar product of these two magnitudes is a time-even pseudoscalar, which is not invariant and changes sign under space inversion ($\hat{P}$), and therefore represents a truly chiral influence (Figure 3.7). This is the critical difference between the regular electronic Hamiltonian system, $\hat{H}$, which is invariant with respect to $\hat{P}$. The eigenvalues, or energy, of a system described by the effect of the electromagnetic forces, are invariant with respect to the parity operator, just as in the case of the energies originating from strong interaction or from gravity, which are also invariant with respect to $\hat{P}$. In operator terminology, unlike purely electromagnetic Hamiltonian systems, the parity-violating terms, $\hat{H}_{pv}$, are odd under space inversion, Equation (4.2).

$$\hat{P}\hat{H}_{pv}\hat{P}^{-1} = -\hat{H}_{pv} \tag{4.2}$$

This equation can be seen as an expression of the nonconservation of parity of weak interactions. For enantiomers, the enantiomeric state functions, $\Psi_R \Psi_S$, of a chiral molecule are interconverted by the space inversion operator $\hat{P}$. It then follows that $\hat{H}_{pv}$ shifts the energy of the enantiomeric states in opposite directions, Equation (4.3):

$$\begin{aligned}\langle \Psi_R|\hat{H}_{pv}|\Psi_R\rangle &= \langle \hat{P}\Psi_S|\hat{H}_{pv}|\hat{P}\Psi_S\rangle = \langle \Psi_S|\hat{P}\hat{H}_{pv}\hat{P}^{-1}|\Psi_S\rangle \\ &= -\langle \Psi_S|\hat{H}_{pv}|\Psi_S\rangle = E_{pv}\end{aligned} \tag{4.3}$$

The electroweak interaction in chiral molecules gives rise to a parity-violating shift in the electronic binding energy, E_{pv}, which is positive for one enantiomer and negative for its mirror image. Hence, the energy difference between different chiral states in a molecule is given by Equation (4.4):

$$\Delta E_{pv} = 2|\langle \Psi|\hat{H}_{pv}|\Psi\rangle| = 2E_{pv} \tag{4.4}$$

From a historical perspective, it was suggested in the early 1960s that parity violation could be the origin of biological chirality, determining the preference of the enantiomers chosen by Nature, e.g., L-aminoacids or D-sugars.[14,34] Quantitative *ab initio* calculations of parity violation for molecules were developed[35] and were soon applied to the building blocks of life, yielding preliminary and perhaps overenthusiastically greeted results.[36] The conclusions of all this work were that: (a) ΔE_{pv} was very small, of the order of 10^{-20} to 10^{-17} hartrees, and (b) in nearly all cases the natural enantiomer was found to be the more stable.[37] Calculations displayed preference of L- over D-amino acids and D- over L-sugars, and this was also true for the secondary natural chiral structures of the α–helix and the β–sheet in proteins, the right-handed helical DNA (which is its natural helicity) and thiosubstituted analogs, which produced even higher ΔE_{pv} values due to the presence of heavier atoms.

More recently, this series of results has proven to be very controversial and has been contested for a number of reasons.[38,39] In fact, calculation of ΔE_{pv} is a very difficult problem and requires a cautious approach. Modern theory treats parity-violating potentials as tensors in which the individual elements E_{pv}^{xx}, E_{pv}^{yy}, E_{pv}^{zz} may have different signs and evolve independently in the calculations, leaving the sign and magnitude of former calculations of ΔE_{pv} uncertain and rendering the theory prior to 1995 obsolete.[40] The implementation of these more realistic theories, and methods such as configuration interaction,[41,42] and consideration of relativistic effects,[43] and technical improvements in computers, have provided more reliable data. A reasonable degree of convergence between different groups of research using independent calculations has been attained, which provides a way—and so far the only way—to test the consistency and reliability of calculation of ΔE_{pv}.[40] On these grounds, more rigorous calculations of the parity-violation energy difference between enantiomers of biologically relevant molecules may be summarized as follows:

- ΔE_{pv} is between one and two orders of magnitude larger than previously accepted. It is still very small, however, roughly of the order of 10^{-19} to 10^{-17} hartrees (i.e., 10^{-16} to $10^{-14} kT$ at 25 °C, equivalent to ca. 10^{-17} to 10^{-15} kcal mol^{-1}). At room temperature this corresponds to an enantiomeric excess of $ee = \tanh(\Delta E_{pv}/2kT) \cong E_{pv}/kT = 10^{-16}$ to 10^{-14}, i.e., a tiny excess, amounting to ca. 10^6–10^8 molecules per mol of the energetically favored enantiomer under thermodynamic equilibrium at 25 °C.

- Earlier conclusions on the relative stability of L- compared with D-α–amino acids, as well as other biomolecules, fall apart due to the large errors inherent in the older theory. At the current state of ΔE_{pv} calculations, a generalized stabilization of the naturally occurring enantiomer is not supported.

Reaffirmed and contested several times, the case of alanine is an example of the challenges inherent in these studies. Calculations on alanine and other L-α–amino acids (valine, serine, aspartate) showed that they were indeed more stable their D-mirror images in their zwitterionic form.[44] But other independent investigations have arrived at opposite or inconclusive results.[45] A recent study of L-alanine zwitterions, hydrated with 0–10 molecules of water using the Monte Carlo simulation, has produced an overall destabilization of the L-isomer in a highly scattered set of results.[46]

The parity-violating energy shift is a very complicated function of the molecular conformation. Small geometric variations result in large and unpredictable changes in the value, range and sign of ΔE_{pv}. In water, α-amino acids are present as solvated zwitterions. A large number of exchanging conformations solvated with an undetermined number of molecules of water contribute to the overall energy minimum and overall ΔE_{pv}. An accurate model for the solvated amino acids, even for hydrated L-alanine, seems far from being theoretically available. The sugar precursor, D-glyceraldehyde, in its hydrated form (in aqueous solution) appears also to be more stable than L-glyceraldehyde.[47] In any case, the debate generated concerning the relative stability of some of these biomolecules has been more strident than its uncertain relevance. In the context of a multistep prebiotic synthetic sequence, both far from equilibrium and fuelled by an external source of energy, the prevalence of certain molecules is certainly not necessarily driven by thermodynamic considerations. In spite of the unanswered questions, and bearing in mind that some flaws or fundamental omissions might be present in all current theories, there is a general recognition that a small thermodynamic bias does not necessarily favor certain crucial biological or prebiotic molecules such as L-α-amino acids or D-sugar precursors.[48,49] Nevertheless, the ultimate test for ΔE_{pv} is necessarily experimental.

4.2.5.1 Experimental Tests for Parity-Violating Energies ΔE_{pv} in Molecules

The difference in energy due to the weak interaction ΔE_{pv} should manifest itself through different contributions to the overall thermodynamic energy, and should therefore be observable, for example, by rotational, vibrational or electronic spectroscopic techniques.[50] The current search for such effects includes MW,[51] IR,[51,52] NMR,[53] UV–Vis[54] and Mössbauer[55] spectroscopy, among others.[56] Two qualitatively different kinds of techniques are used. One may broadly distinguish between spectroscopic techniques operating with samples in the gas phase (isolated molecules) and experiments on macroscopic condensed phases (solid or liquid). For the moment, only experiments of the first kind are capable of accurate theoretical treatment, and hence evaluation. *Ab initio* methods and the fundamental theory are not sufficiently developed to be applied to condensed phases and to fully explain their properties. Parity-violating terms for these hypothetical *ab initio* methods are even further away. Any claimed parity-violating effects in macroscopic samples must be treated with extreme caution. Experimentally, they are very likely to be complicated by a number of unwanted effects which are difficult to assess.

A number of questions might arise concerning condensed phases, for instance the optical purity which can be guaranteed for each enantiomer sample, or whether two enantiomerically pure samples are exact mirror-images, and if so, is this true to the 18th figure of the *ee*? Attractive intermolecular forces present in condensed phases are certainly much greater in magnitude than parity-violating effects. Different intermolecular interactions between major and minor enantiomers (diastereomeric interactions) are hard to evaluate, not to mention the unknown influence of minor traces or impurities—and they overcome by far the purely weak-interaction forces. Claims of experimental confirmation of parity-violating effects in condensed phases have been published. Some experiments involve Laser Raman spectra or solid NMR of enantiomeric amino acids,[57] or preferential crystallization of transition metal-containing complexes.[58] It is generally agreed that these should be treated with caution for the time being. Often, these claims have later been discarded as flawed by the scientific community[59] (see also Section 5.5, *The Salam Phase Transition*).

Rotovibrational spectroscopy (MW, IR) can be performed in the gas phase and therefore analyzed as isolated molecules. Energy differences due to parity violation in the ground and excited states of enantiomer molecules should, at least theoretically, be noticeable as a small difference in their absorption frequency, Δv. In this context, halogenomethanes have been chosen for this challenging purpose. Examples such as CHBrClF, which are certainly among the simplest chiral molecules, are good models for the study of fundamental chirality. They have attracted interest due to a number of features which make them ideal candidates for parity-violating experiments and calculations. They are simple, volatile, contain heavy atoms and are devoid of conformational complications which might overlap and obscure spectra.

Bromochlorofluoromethane itself has been the subject of intense theoretical[60] and experimental research.[61] Enantiomerically enriched CHBrClF has been synthesized and partially resolved samples submitted to spectroscopic analysis. One of the highest level of spectroscopic precision attainable by experiment is found in the IR domain, and can be achieved by means of IR laser excitation. Using a high-resolution CO_2 laser with emission range around $1000\,cm^{-1}$, an ultra-high resolution infrared spectroscopic analysis of the C–F stretching fundamental v_4 of CHBrClF ($v_{(exp)} = 1077.2\,cm^{-1}$) was carried out, giving precision down to the level $\Delta v/v_{(exp)} = 10^{-13}$.[61] In parallel, calculations have been carried out for the same fundamental vibration by a number of groups. They have established the actual effect of parity-violating frequency differences between enantiomers, $\Delta v/v_{(calc)} = 10^{-17}$, some 3 to 4 orders of magnitude lower than the smallest Δv experimentally detectable at present.

Calculations, together with spectroscopic properties, can provide a complete description of the molecular physics and dynamics. From the molecular partition function, incorporating the parity-violating terms with the translational, rotational, vibrational and electronic elements, a complete range of thermodynamic state functions (Gibbs energy, G, enthalpy, H, *etc.*) can be obtained for a given temperature in the gas phase. In Figure 4.9 some representative thermodynamic data of a pair of enantiomers are given as an example. The enthalpy of reaction at 0 K for the inversion of configuration of the enantiomers, (S)-CHBrClF ⇌ (R)-CHBrClF, is $\Delta H_0° \approx N_A \Delta E_{pv} = 5.478 \times 10^{-15}\,kcal\,mol^{-1}$, where N_A is Avogadro's number. At 300 K, the equilibrium constant for the same reaction, $K = 0.999\,999\,999\,999\,990\,316$, corresponds to a positive free energy, $\Delta G° = 5.769 \times 10^{-15}\,kcal\,mol^{-1}$. The equilibrium is shifted towards (S)-CHBrClF, which is more stable than (R)-CHBrClF, in agreement with the sign of $\Delta H_0°$ or ΔE_{pv}.[62] The enantiomeric excess of this thermodynamic equilibrium is exceedingly small, $ee = (1-K)/(1+K) = 4.842 \times 10^{-15}$, but is still not quite zero.

In the literature it is possible to find both proposals and actual attempts to observe molecular parity violation through spectroscopic means. So far all have

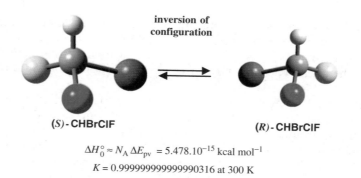

$\Delta H_0^\circ \approx N_A \Delta E_{pv} = 5.478 \cdot 10^{-15}$ kcal mol^{-1}
$K = 0.9999999999999990316$ at 300 K

Figure 4.9 Bromochlorofluoromethane is one of the simplest chiral molecules available. It has been the subject of intense research, both theoretical and experimental, concerning parity-violating differences between S and R enantiomers. According to recent calculations, ΔE_{pv} favors the S enantiomer by 8.736×10^{-18} hartree (5.478×10^{-15} kcal mol^{-1}). At temperatures close to ambient (300 K), the hypothetical thermodynamic equilibrium $S \leftrightarrows R$ is shifted from unity ($K = 1$) by a small excess of the more stable enantiomer, ca. 2.916×10^9 molecules per mol. Currently, experimental validation of molecular parity violation for enantiomers in any of its forms is still missing, and is being actively sought.

either failed, or at best have been unable to confirm it beyond doubt. In recent *ab initio* relativistic calculations, large parity violation effects in heavy metal-containing chiral compounds have been proposed.[63] By careful selection of ligands around the heavy metal center, most peripheral atoms can produce additive (or at least noncanceling) parity-violating effects. The largest effects have been reported in $Os(\eta^5\text{-}C_5H_5)(=CCl_2)(PH_3)Cl$ ($\Delta E_{pv} = 1.50 \times 10^{-11}$ kcal mol^{-1}), and $Re(\eta^5\text{-}C_5H_5^*)(=O)(CH_3)Cl$ ($\Delta E_{pv} = 5.18 \times 10^{-11}$ kcal mol^{-1}). These complexes are either real or slightly modified real compounds and show parity-violating differences between enantiomers of $\Delta v/v_{(calc)} = 10^{-13}$ (Δv_{pv} ca. 1 Hz between R and S) in Os = C, (901 cm^{-1}) and Re = O, (989 cm^{-1}) stretching modes. In the view of the present authors, the issue should become accessible to high-resolution vibrational spectroscopy using a tunable CO_2 laser.[63a] For the moment, however, the observation of a parity-violating effect induced by weak neutral currents in enantiomeric molecules by spectroscopic means is only a dream. But it is the object of intense research, and no doubt some light will be shed on the subject before long.

4.2.6 β-Radiolysis

The underlying nature of β-radioactive decay—already described in detail—is entirely governed by the weak force (see Section 4.2.2, *The β–Decay*, Figure 4.5). β-Decay electrons, or β-rays, have helicity of −1, and are therefore inherently chiral in the laboratory frame (this is the case for this highly energetic radiation, traveling at close to the speed of light, since chirality depends on the velocity of the observer). More accurately, these electrons have a velocity-dependent chirality which increases with v/c.[64] The link between this source of chirality and the origins of biological chirality was proposed shortly after the discovery of parity violation.[65] As β-radiation travels through matter it slows down, with concomitant emission of electromagnetic radiation. This is the so-called "Bremsstrahlung radiation" (braking radiation), which is circularly polarized in the same sense as the original β-radiation. This radiation covers continuously the entire spectrum of energy, having a maximum degree of polarization at the highest energies (γ-rays) and decreasing progressively at lower frequencies.

F. Vester and T. L. V. Ulbrich suggested that the asymmetric photochemistry carried out by circularly polarized Bremsstrahlung radiation could be the origin of enantiomeric excess in prebiotic molecules. Also based on the weak force, this is a completely different approach to the previous search for parity-violating mechanisms, based on thermodynamic imbalances in the energy of the enantiomers. However, attempts to confirm this hypothesis in the laboratory were unsuccessful.[66] Over the following decades, great effort was applied to substantiate the hypothesis using many different isotopic β-emitters over a wide range of experimental conditions. While some positive claims have been sporadically reported, lack of reproducibility, and in most cases lack of evidence of enantioselective β-radiolysis, have precluded experimental validation of this hypothesis.[67] Despite these disappointing results, it is fair to remark that the β-decay hypothesis represents a serious candidate for the initial source of chirality, given its univocal, well defined chiral sense. The β-decay is by no means a rare phenomenon on Earth, and is closely linked to natural radioactivity. An estimate of geothermal energy—mainly provided by radioactive heating in the mantle and crust—amounts to a heating power of some 24–36 TW,[68] which gives some idea of the magnitude of the phenomenon. The most abundant radioactive isotopes in the Earth's crust, ^{238}U, ^{232}Th and ^{40}K, which are mainly responsible for this heating, decay through a more or less elaborate radioactive series involving a number of β-emission steps. The main drawback of this mechanism lies in the smallness of the number of the particles decaying in comparison to the size of a macroscopic sample. From those radioactive isotopes, ^{40}K may have been among the strongest overall radioctive influences (half-life 1.248×10^9 years). Natural potassium represents the seventh most abundant element in the Earth's crust (ca. 2.6% by weight in the upper crust), and this isotope constitutes 0.0117% of the natural K and is a β-emitter (88.8% of atoms β-decay to ^{40}Ca). Unfortunately, the rate of decay is extremely low, corresponding (from the half-life) to an activity of 31.7 Bq g^{-1} (31.7 disintegrations per second per gram of K, corresponding to 28.2 β-decays g^{-1}s^{-1}). Some 3800 million years ago—the estimated date of the appearance of life on Earth—the level of radioactivity was necessarily higher, but still extremely small, ca. 262 Bq g^{-1} for this isotope. For more concentred sources of radioisotopes (e.g., a vein of uraninite), or for more radioactive isotopes that might have existed in increased concentration at that time, the situation is not very different considering the overall estimated age of the Earth (ca. 4500 million years) and the faster decay of the most radioactive isotopes. In a similar fashion any detectable amount of chiral products obtained in the laboratory is unlikely, unless efficient mechanisms of amplification are operative. In spite of the experimental effort devoted to this study, the conditions under which chirality at the elementary particle level may be transmitted to the molecular level by this mechanism remain unknown.

4.3 Asymmetric Photolysis and Photosynthesis

Natural and magnetochiral circular dichroism may in principle cause an enantiomeric excess in a racemic mixture, or in the reaction products from an achiral starting material, when irradiated by circularly polarized UV light or by nonpolarized UV light in the presence of a magnetic field collinear to the light beam, respectively. In either of these cases, these processes are described as asymmetric photolysis or asymmetric photosynthesis, depending on whether the residue of one enantiomer or the reaction products, respectively, are of relevance in the process.

4.3.1 Circularly Polarized Light: Circular Dichroism

Circularly polarized light (CPL) is an example of a truly chiral influence, in the sense defined by Barron (see Chapter 3, *The Concept of Chirality*, and Section 3.3.3, *Experiments under True Chiral Influence*) and as such it should be capable of performing asymmetric photochemistry. In common with all chiral entities, CPL exists in pairs of opposite helical sense, (r)-CPL and (l)-CPL, which can occur either alone or superimposed with the same phase and intensity in plane-polarized light. Due to the intrinsic chirality of CPL, the interaction with a chiral or a chirally perturbed chromophore is not equivalent for (r)-CPL and for (l)-CPL, resulting in differential absorption (or molar extinction coefficient, ε) of right- and left-circularly polarized light ($\Delta\varepsilon = \varepsilon_r - \varepsilon_l \neq 0$) and emergence of the phenomenon of circular dichroism (CD). If we change the roles of the radiation and the substrate, *i.e.*, we start with a racemic chromophore (such as a mixture of D- and L-enantiomers) and CPL of one handedness, the experiment is conceptually equivalent, producing an identical differential of molar extinction coefficient ($\Delta\varepsilon = \varepsilon_D - \varepsilon_L \neq 0$), but becomes much more attractive for our purposes.

Starting from nonchiral starting materials, such as achiral molecules or racemic mixtures, chiral mixtures of final products can be obtained using CPL through an asymmetric photochemical process. These processes can follow either photodegradative or photosynthetic pathways. In either case, the starting materials must absorb radiation in the visible or UV region, like any other photochemical reaction. However, CPL is preferentially absorbed by one enantiomer, which therefore suffers greater photochemical processing than the other enantiomer. The result is the enrichment of the more transparent enantiomer in a type of asymmetric photodegradation which generates enantiomerically enriched mixtures from racemates under the sole influence of CPL. On the other hand, in an asymmetric photosynthesis it is possible to start from an achiral substrate (*i.e.*, one having any S_n symmetry element, with n = 1, 2,...), which upon irradiation with CPL is converted preferentially to a chiral excited state, eventually leading to a favored enantiomer. In addition, we can talk of asymmetric photoisomerization when chiral molecules undergo enantioselective isomerization induced by CPL (*e.g.*, the interconversion between enantiomers $R \leftrightarrows S$). This is nonetheless the result of preferential absorption by one enantiomer, just as in the case of asymmetric photodegradation, with evolution towards the enrichment of the opposite enantiomer. A summary of asymmetric photochemistry induced by CPL is shown in Scheme 4.1.

Examples of asymmetric photodegradation have been observed in many instances, whereas asymmetric photosynthesis is rare.[69] Asymmetric photodegradation with CPL is one of the most important mechanisms for the abiotic generation of enantiomerically enriched compounds.[70] Asymmetric photodegradation has a wide range of applications, virtually any type of racemic compound possessing a chromophore being a candidate. The enantiomeric excess attainable depends on the difference in molar extinction coefficients for the two enantiomers ($\Delta\varepsilon = \varepsilon_D - \varepsilon_L$) relative to the average extinction coefficient [$\varepsilon = 1/2\,(\varepsilon_D + \varepsilon_L)$], or *anisotropy factor*, $g = \Delta\varepsilon/\varepsilon$, and the extent of the photodegradation. The greater the conversion (tending to 100%), the higher the enantiomeric excess obtained (also tending to 100%), but at the expense of a lower yield of the remaining optically enriched enantiomer (which tends to 0%). The reactions shown in Scheme 4.2 are an example of this. Asymmetric photolysis of racemic leucine yields enantiomerically enriched leucine using laser-induced CPL in the UV region (213 nm, n → π* electronic transition), the sign of rotation depending on the sign of the CPL.[71,72] The enantiomeric excess is 2%

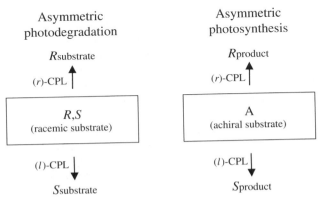

Scheme 4.1 Circularly polarized light is a truly chiral electromagnetic radiation which can induce asymmetric photosynthesis, in addition to asymmetric photodegradations starting from nonchiral systems. The signs shown for CPL are arbitrary.

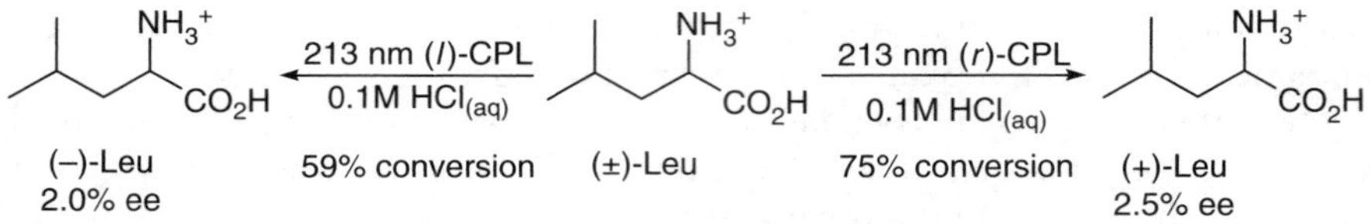

(−)-Leu ⟵ 213 nm (*l*)-CPL, 0.1M HCl$_{(aq)}$ ⟵ (±)-Leu ⟶ 213 nm (*r*)-CPL, 0.1M HCl$_{(aq)}$ ⟶ (+)-Leu
2.0% ee   59% conversion   75% conversion   2.5% ee

Conversion	Enantiomeric Excess
59%	2%
75%	2.5%
93.7%	*5.0%*
98.4%	*7.5%*
99.6%	*10%*
99.9%	*12.5%*
99.975%	*15%*

Scheme 4.2 Asymmetric photodegradation of racemic leucine (Leu). (*r*)-CPL and (*l*)-CPL give (+)-Leu and (−)-Leu, respectively, with increasing enantiomeric excess as the conversion approaches 100% (figures in italics are extrapolated).

at 59% conversion of the initial leucine, increasing to 2.5% at 75% conversion. Extrapolation of *ee* at higher conversions, such as those in italics in Scheme 4.2, can easily be calculated. The enantiomeric excesses obtained in the laboratory in the condensed phase have similar values to the enantiomeric excesses detected in meteoritic amino acids, but no CD effect has been observed in the gas phase.[73] Recently, similar values of enantiomeric excess for D-leucine (+2.6%) have been reproduced under simulated interstellar/circumstellar conditions, when irradiating with (*r*)-CPL at 182 nm, near to the $\pi \to \pi^*$ electronic transition.[74]

Considering now the second scenario in Scheme 4.1, asymmetric photosynthesis, the most representative example of this is probably the case of the helicenes (Scheme 4.3). Circularly polarized light will preferentially excite the ground-state molecules. The ground state is achiral at a first approximation, but preferential absorption of CPL by chiral conformations is a possible alternative mechanism. This will give rise to an excited state with a certain helical sense, which leads to enantiomerically enriched products of defined chirality. Helicenes have an extremely intense optical rotatory power (*e.g.*, for hexahelicene (**5**), $[\alpha]_D^{24} = 3640$) and can be produced from flat precursors, the only chiral influence arising from CPL. These are examples of molecules in which the chromophore itself is chiral—the π-system—and the chromophore extends over the entire molecule. Very low enantiomeric excesses in helicenes can be directly measured by polarimetry. Thus, 1-(2-benzo[*c*]phenanthryl)-2-phenylethylene (**3**) or 1-(2-naphthyl)-2-(3-phenanthryl)-ethylene (**4**) were irradiated with CPL in the wavelength range 310–410 nm in the presence of iodine as an oxidant, to give hexahelicene (**5**) *via* a dihydrohelicene intermediate.[75,76] The optical rotation of the newly synthesized *M*-hexahelicene was $[\alpha]_{436}^{23} = +30.5$ using (*l*)-CPL as a chiral activator, whereas the mirror-image experiment afforded *P*-hexahelicene, $[\alpha]_{436}^{23} = -30.0$ by irradiation with (*r*)-CPL, starting in both cases from **4**. In contrast to asymmetric photodegradation, asymmetric photosynthesis has received little attention as a natural source of chiral molecules.

Chiral Physical Forces

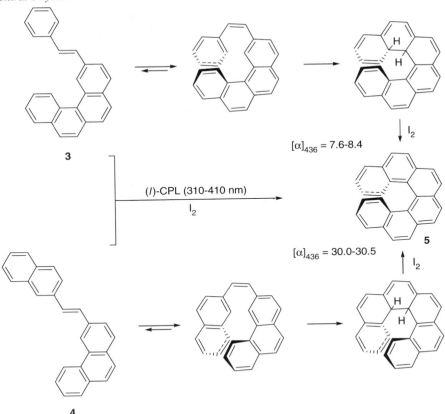

Scheme 4.3 Asymmetric photosynthesis of hexahelicene (5) from planar 1,2-diaryl ethenes. Both syntheses afford slightly enriched (+)-hexahelicene when (*l*)-CPL is used. The mirror-image experiment, using (*r*)-CPL, yields (−)-hexahelicene with similar optical purity, within experimental error.

One effect which arises when dealing with asymmetric photochemistry is the Khun–Condom sum rule. According to this, bands of circular dichroism alternate in sign and add up to zero over the whole spectrum. Circular dichroism of L-leucine, for instance, shows a positive CD band for the [π^*, n] transition (211 nm), a negative CD band for its [π^*, π_1]-transition (183 nm), and probably a second maximum (positive) for the [π^*, π_2]-transition (142 nm).[77] From the definition of molar ellipticity, $\Theta = 3300\,(\varepsilon_l - \varepsilon_r)$, it can be inferred that the [π^*, n]-transition of L-leucine is caused preferentially by left-circular polarized light (*l*-CPL), the [π^*, π_1]-electronic transition of L-leucine by *r*-CPL, and so on. Thus, starting from racemic leucine, at the 211 nm [π^*, n]-transition *l*-CPL is assumed to result in an enantiomeric excess of D-leucine, at the 183 nm [π^*, π_1]-transition *l*-CPL it should lead to an *ee* of L-leucine, and so on, the total effect being theoretically zero over the entire spectrum. In practice, things are not so strict and mathematical restrictions can sometimes be bypassed. For instance, in the case of a molecule in which only CD bands of the same sign trigger photochemical reactions, whereas bands with the opposite sign result in dissipative absorption, asymmetric photodegradation is perfectly possible. In any case, the Kuhn–Condom sum rule poses problems for the preferential photodegradation of one enantiomer and for asymmetric photosynthesis, unless a narrow or monochromatic CPL band is employed.

4.3.1.1 *Circularly Polarized Light on Earth*

Natural light, in common with all common sources of incandescent light, including sunlight and flames, is unpolarized. However, under certain conditions small amounts of CPL are produced naturally on Earth. A small degree of circular polarization (<0.5%) has been reported in solar radiation at twilight, due to multiple aerosol scattering in the atmosphere.[78] However, the Earth's rotation causes CPL of opposite polarity to occur at dawn and dusk, averaging zero over the whole sky and over the diurnal cycle, and it is therefore necessary to propose supplementary influences to obtain the asymmetry necessary to provide enantioselective photolysis in an initial racemic mixture of chiral molecules. This might be achieved by interrupting either dawn or dusk natural light by means of a landform, *e.g.*, by a mountain or a cliff oriented east or west of the photolysis spot, leading to unequal exposure. However, such east- or west-oriented local topography is likely to be equally distributed on Earth. On the other hand, reversed temperature profiles in the oceans could provoke destruction of, say, the D-isomer by CPL at dusk in a sea surface hotter than at dawn, creating a daily L-isomer excess protected from radiation by nightfall, and preserved by diffusive down-flow into cold, darker regions, with an eventual accumulation of the L-amino-acid excess in the depth of the oceans.[79,80] Another source of CPL is provided by the Faraday effect, resulting from the interaction of sunlight with the terrestrial magnetic field. This effect has the opposite sign in each hemisphere and is mostly canceled out throughout the yearly cycle, but it does not average to zero due to the obliqueness of the ecliptic and the elliptical nature of the Earth's orbit around the Sun.[81] Other small sources of CPL have also been reported.[82] The main drawback common to all these mechanisms is the weakness of the resulting circular polarization in an otherwise polychromatic solar radiation. The effect of the small anisotropy factors expected ($g = \Delta\varepsilon/\varepsilon$) may be further depleted as a result of the Khun–Condom sum rule, resulting in a vanishingly small net effect unlikely to give rise to enantioselective photochemistry.

4.3.1.2 *Circularly Polarized Light in Outer Space*

The interstellar medium—the regions of space between the stars—is occupied by a very dilute gas, mostly consisting of hydrogen. There are regions of relatively high density of matter, called interstellar clouds. Planetary systems, including the solar system, are believed to have formed from the gravitational collapse of matter in these clouds. Temperatures in the interstellar clouds are low (a few tens of K) and all but the most volatile species (H_2, He) is solid and condenses on to dust particles, forming "grains". Grains accrete ice layers composed primarily of water, but also contain CO, CO_2, CH_3OH, and NH_3 as the main molecular components.[83] The ice can undergo considerable processing by stellar UV radiation.[84] Many of the organic molecules present in carbonaceous chondrites (carbon-containing meteorites, see above), comets and asteroid dust are thought to arise, at least in part, from the ice and complex compounds synthesized in the interstellar medium, which later undergo more or less intense processing.

Experimental confirmation of this was recently provided in the laboratory.[85,86] A variety of amino acids and other organic molecules were synthesized from C_1 precursors such as CO, CO_2, CH_3OH and other simple molecules such as NH_3, in an ice (H_2O) matrix mimicking interstellar conditions (12–15 K, 10^{-7} to $^{-8}$ mbar) by UV irradiation (7.3–10.5 eV, ca. 170–120 nm; molecular transitions are centered at ca. 160 nm). Glycine, alanine, sarcosine, β-alanine, β-aminobutyric acid, valine, proline, serine, aspartic acid and other non-proteinogenic amino acids, *inter alia,* were obtained in racemic form by irradiation of ice with unpolarized UV light as activation energy. The asymmetric version of this experiment has not been reported until very recently.[87] Irradiation with UV circularly polarized light (UV CPL, 167 nm) of a laboratory sample of interstellar ice analogs under the interstellar conditions described above provided, after careful elimination of statistical and systematic uncertainties, no discernible enantiomeric excesses in the amino acids studied (<1% *ee* in alanine and 2,3-diaminopropanoic acid).

In space, large amounts of circularly polarized light were first detected in a region of the Great Nebula in Orion, known as Orion Molecular Cloud 1 (OMC-1, Figure 4.10).[88] This nebula belongs to what astronomers call an H II region, a cloud of glowing gas and plasma in which intense activity of new star formation is taking place and organic molecules are abundant, receiving its name from the large amount of ionized atomic hydrogen (or H II) these contain. The largest degrees of circular polarization observed in OMC-1 were at near-infrared wavelengths (2.2 μm). These are thought to arise from the scattering of stellar radiation from aligned dust grains. Unfortunately, the obscuring effect of the dust prevents these regions being observed directly at shorter wavelengths, at which photochemistry might be feasible. The CD bands in amino acids are found in the UV at wavelengths around 200–230 nm (Scheme 4.2). Nevertheless, calculation indicates that scattering should also produce comparable degrees of circular polarization extending down to UV wavelengths. The sign of the circular polarization encountered by such molecules will depend on the local scattering geometry, and is in principle equally likely to be (*l*)- or (*r*)-CPL in different regions. Over the entire universe these chiral effects are canceled out, although their local effects are so large that they might involve areas the size of entire solar systems. High degrees of circular

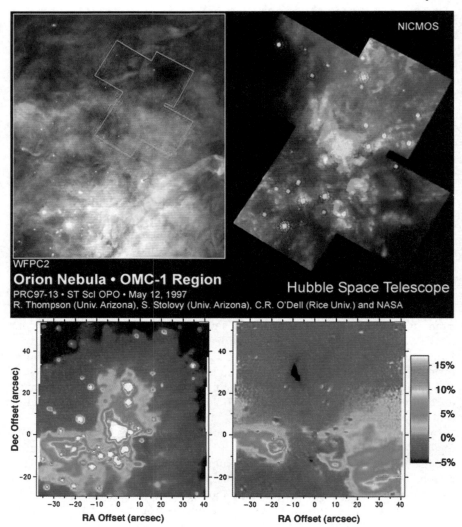

Figure 4.10 OMC-1 is a giant molecular cloud in the Orion nebula, located in the "sword" of the constellation Orion (*upper left*). The nebula harbors a region of our galaxy in which new stars are being formed. Such regions are known to be rich in organic molecules. Most of the objects in this region are invisible, because optical light is blocked by the interstellar dust trapped in OMC-1. Infrared light, however, penetrates the dust to uncover the very heart of OMC-1 (*upper right* and *lower left*). In this region strong infrared circular polarization resulting from dust scattering has been observed, its percentage being represented in the lower right picture (as high as 17%, in the red/white areas). (Pictures from http://hubblesite.org, and from Reference 88, with authorization.)

polarization have been also found in the star-forming complex NGC 6334, in the constellation Scorpius. NGC 6334 is another giant H II region and molecular cloud similar to Orion.[89] These two areas of major circular polarization suggest that the conditions needed for selective photolysis by circular polarization may not be a rarity in regions of massive star formation. Other possible astronomical sources of circularly polarized UV light have been pointed out, for example the Rubenstein–Bonner neutron star hypothesis.[90] In this hypothesis, synchrotron radiation from magnetic neutron stars is suggested as a possible source of UV CPL.[91] However, it is not predicted to be strongly circularly polarized, and most importantly, it has never actually been observed.[89] Magnetic white dwarfs and white dwarf binaries (Polars) can be highly circularly polarized but any effect on molecular clouds and star formation regions must rely on rare chance encounters.[89] Another drawback of these sources is that photochemical enantiodiscrimination works better with monochromatic light tuned to the wavelength of a major CD band maximum. In either of the former cases, the expected band-widths covers the entire UV spectrum, thus in principle diminishing[92] the efficacy of the process.[93]

4.3.1.2.1 Panspermia. CPL from the scattering of star light is a very large chiral effect in Nature, and new theories have arisen from it. It could be connected to the origins of homochirality on Earth by different routes. J. Oró proposed long time ago that the delivery of extraterrestrial organic molecules may have been crucial for the origin and early evolution of life.[94] This is consistent with current evidence for the earliest life on Earth having arisen about 3800 million years ago, coincident with the end of the heavy bombardment phase. Comets, asteroids or interplanetary dust containing organic chiral molecules could have originated in a region such as OMC-1 and might have provided the material necessary for an initial enantiomeric enrichment *via* meteorites.[95] The enantiomeric excess found in Murchison (up to 9% of L-enantiomer) is about as high as can reasonably be expected from the action of UV CPL (see Chapter 7, *Outside Earth: Meteorites and Comets*), and must have required conditions such as those found in OMC-1, which are not very common in space. Also, the formidable journey necessary to deliver the chiral load severely reduces its chances of reaching the target, Earth. The question moves now one step backwards, and relies on a more profound question yet to be answered: is life commonly present in the universe?

4.3.1.2.2 Was Our Entire Solar Nebula Under the Influence of CPL? The Solar System is believed to have formed according to the nebular hypothesis, which entails emergence from the gravitational collapse of a giant molecular cloud about 4600 million years ago. In a different hypothesis to the panspermia, the OMC-1 region might also have been similar to the region in which our own solar system was formed. The discovery of high degrees of circular polarization provides in principle a realistic mechanism for explaining the origin of chirality, not only in meteorites but also on Earth. The extent of the region where CPL of single-handedness originates is very large.

The CPL in such a region could imprint a preferred handedness on any organic molecules in the region, including those in a cloud beginning to collapse to form a star and its planets. It is likely that the whole of a planetary system would thus acquire a single handedness of CPL. From this we can speculate that chirality in meteorites and on Earth, including ourselves, would have a similar origin. We are, after all, simply stardust.

Asymmetric photolysis of amino acids by stellar CPL is in fact one of the most solid arguments we have which could explain the enantiomeric excess found in meteorites. On Earth, however, rupture of chiral symmetry is only found in biomolecules and related superstructures. Apart from that, no abiotic remnant of that primeval chirality is found. Most theories are not without important drawbacks. The origin of life is inevitably linked to chemistry in the aqueous phase. Racemization is an inexorable process. It can be circumvented in cold/dry stages in the thermal/processing history of interstellar molecules, e.g., during their transit to Earth,[96] but this is less so within the Earth's crust, the majority of which is covered with water. The racemization lifetime of amino acids under wet conditions has been estimated to be only about 10^4 to 10^6 years,[97] a very short space of time.

Asymmetric photochemistry involving π-extended aromatic hydrocarbons should also be regarded as a potential pathway mediating the transfer of chirality to organic molecules. Carbonaceous chondrites (see Chapter 7, *Outside Earth: Meteorites and Comets*) contain large amounts of organic carbon. As much as 70–80% of the total carbon is present as a carbonaceous matrix of poorly understood structure.[98] In the presence of CPL, vast π-extended systems can behave as traps of very effective cross-section for CPL photons. In fact, polycyclic aromatic hydrocarbons (PAHs) are characterized by very large extinction coefficients (ε), several orders of magnitude larger that the carboxyl chromophore in amino acids. PAHs are abundant in the interstellar medium,[99] being also structural constituents of the macromolecular carbonaceous matrices such those found in meteorites.[100] Due to their widespread occurrence, PAHs could well act as efficient traps for the surrounding CPL. Elaborated organic molecules could have been synthesized after a chiral activation, originated by CPL, and delivered by a macromolecular π-extended system. Once photoactivated, two possible productive outcomes can be envisaged. Firstly, chiral intramolecular reactions, including chiral folding, can occur on the lines of the reactions described in Scheme 4.3 for helicene. Chiral PAHs obtained this way could later serve as matrices for further chiral chemistry or photochemistry, e.g., on their surfaces, to yield elaborated organic molecules. The adsorptive character of the charcoal-like surface of many organic molecules is well known. Secondly, chiral PAHs offer another route for passing on activation energy and chirality directly to other simple interstellar molecules, precursors of more complex organic compounds, in a type of asymmetric photosensitization. Either standpoint is for the moment only hypothetical and will require experimental validation.

The Kuhn–Condom sum rule clearly restricts the efficacy of astronomical sources of CPL in photochemical reactions. This rule for rotational strength may limit the efficiency of a preferential photodegradation of one enantiomer, and also of asymmetric photosynthesis, unless nearly monochromatic CPL radiation is employed. These sources of radiation, if any, are scarce in the Universe, although it has been pointed out that the UV spectra of most stars are uneven in intensity over the full range of wavelengths considered.[88] Finally, only a small proportion of stars may have formed in a suitable polarization environment to develop the chiral asymmetry needed for life. Again, the unanswered question of how frequently life is found in the Universe is an underlying factor in these hypotheses.

4.3.2 The Magnetochiral Effect

Magnetochiral birefringence and magnetochiral dichroism, related to differences in light refraction and light absorption, respectively, in racemic media under the effect of a magnetic and an unpolarized electromagnetic field, have been theoretical established for a long time as a link between chirality and magnetic fields.[101] Such a link could discriminate between the two enantiomers of a chiral system.[102,103] Regarding the second effect, which is the only one of importance for the purpose of enantiomeric enrichment, it is predicted that a static magnetic field, $\vec{B}$, parallel to the direction of propagation of an incident nonpolarized light beam, $\vec{k}$, causes a small shift in the value of the absorption coefficient (or extinction coefficients) of a chiral molecule, which is opposite in sign for enantiomers, and can be reversed by aligning $\vec{B}$ and $\vec{k}$ in an antiparallel manner. Indeed, the combined effect of the two, corresponding to the scalar product $\vec{k} \cdot \vec{B}$, is a *truly chiral influence*, and has been presented in Section 3.3.3, *Experiments under True Chiral Influence*, and Figure 3.6. Although it was suggested some time ago,[102,103] the substantiation of such a true chiral field with experimental evidence of its effects has only recently been achieved. This is partly because the magnetochiral effect, either in the directional birefringence or dichroism modality for nonpolarized light, is so small that it has been difficult to detect experimentally. In the first magnetochiral dichroism experiment, unpolarized light (Hg lamp) produced excitation of a racemic mixture of Eu(($\pm$)tfc)$_3$ [europium(III) *tris*-(3-trifluoroacetyl-($\pm$)-camphorate)], when luminescence was collected in the directions parallel and antiparallel to the applied magnetic field. A small difference in intensity of emission between the direction of the two magnetic fields was detected.[104] But it was not until 2000 that an actual *ee* was generated by means of the magnetochiral effect. G. L. J. A. Rikken and E. Raupach reported obtaining a measurable *ee* in the [Cr(ox)$_3$]$^{3-}$ complex [chromium(III) *tris*-oxalate],[105] when this enantiomeric mixture was asymmetrically photolyzed by an unpolarized laser beam (100 mW Ti:sapphire laser, near 696 nm) placed under a powerful static and collinear magnetic field (up to 15 T). This chromium complex exists in a Δ- and a Λ-configuration, which are the enantiomeric right- and left-handed versions of the complex (Scheme 4.4). These complexes are labile in solution and spontaneously dissociate and re-associate. At equilibrium, [Cr(ox)$_3$]$^{3-}$ naturally exists as a

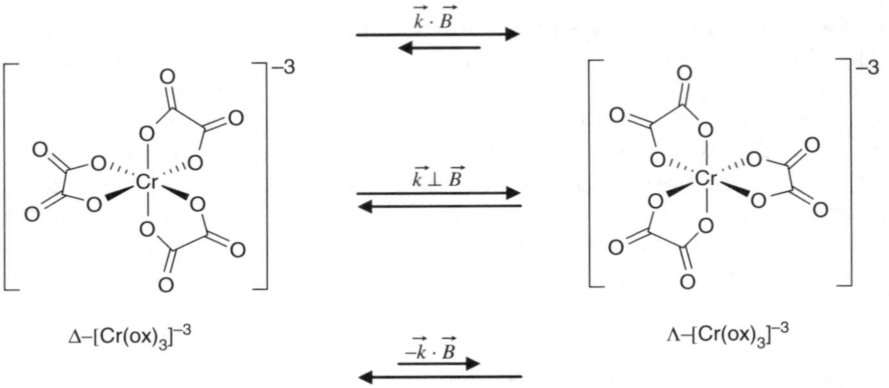

Scheme 4.4 The $[Cr(ox)_3]^{3-}$ complex exists as a racemic mixture of its components, Δ- and a Λ- $[Cr(ox)_3]^{3-}$ in equilibrium. In the magnetochiral dichroism experiment, this equilibrium is shifted by a enantioselective photodissociation of one enantiomer (due to differential absorption of light) by means of a nonpolarized light beam, $\vec{k}$, in the presence of a collinear magnetic field, $\vec{B}$. A stationary *ee* is eventually reached, the magnitude of which is very small, about 1.0×10^{-5} for a magnetic field of 1 T, and is linear with the scalar $\vec{k} \cdot \vec{B}$ product, meaning that the effect is reversed for parallel ($\vec{k} \cdot \vec{B}$) and antiparallel ($-\vec{k} \cdot \vec{B}$) alignment of the light beam and magnetic field, and zero for the perpendicular $\vec{k} \perp \vec{B}$ orientation.

racemic mixture of Δ- and Λ-$[Cr(ox)_3]^{3-}$. The dissociation is accelerated by the absorption of light, so under $\vec{k} \cdot \vec{B}$ irradiation, the more absorbing enantiomer dissociates more rapidly, whereas the subsequent re-association yields equal amounts of the two enantiomers. This leads to an enantiomeric excess of the less absorbing enantiomer, the handedness of which depends on the sign of the scalar $\vec{k} \cdot \vec{B}$ product, *i.e.*, the relative orientation of the vectors. As soon as the irradiation stops, the system returns to the racemic state owing to thermal dissociation and random re-association of the complexes. It was confirmed, as predicted, that opposite chiral effects were observed when the laser beam, $\vec{k}$, and magnetic field, $\vec{B}$, were parallel or antiparallel, respectively, while no effect was measured for perpendicular vectors, $\vec{k} \perp \vec{B}$ (Scheme 4.4). These experiments were followed shortly by experimental detection of magnetochiral birefringence, a difference in CPL light refraction—in place of light absorption—conceptually intimately interrelated.[106] From these experiments, it is proven that a magnetochiral influence ($\vec{k} \cdot \vec{B}$ is a pseudoscalar) possesses true chirality and therefore has the same status as circularly polarized light in its ability to induce absolute enantioselection.[107]

In Scheme 4.4, the magnitude of the *ee* (determined by circular dichroism) is very small, about $ee/\vec{B} = 1.0 \times 10^{-5}\,T^{-1}$, and is linear with the scalar product $\vec{k} \cdot \vec{B}$, indicating that the effect could be greater in the presence of stronger magnetic fields. For comparison, the experiment was carried out using magnetic fields of up to 15 T, while the Earth's magnetic field at the surface is only *ca.* 10^{-4} T. A number of scenarios involving magnetochiral dichroism on Earth

Chiral Physical Forces

have been reviewed.[82] The magnetic field of the Earth in principle provides a source of magnetic field for magnetochiral effects. However, due to the general north–south orientation, the effect would be inoperative in the proximity of the magnetic equator, and also around sunrise and sunset, when the sunlight is perpendicular to the magnetic field. Also, the effect has different signs in either magnetic hemisphere. But the main difficulty arises from the extremely small anisotropy factors associated with magnetochiral dichroism operating within the Earth's magnetic field. These anisotropy factors (g) can be defined by comparing extinction coefficients related to the specific absorption of (+)-CPL or (−)-CPL by the D- or L-enantiomer in an analogous manner to that exemplified in Section 4.3.1, *Circularly Polarized Light: Circular Dichroism*, and are of the order of $g = \Delta\varepsilon/\varepsilon = 10^{-10}$.[82]

In space the situation may be different.[107] Magnetochiral dichroism gains in importance because magnetic fields and unpolarized light are more common than circularly polarized light in the cosmos (Section 4.3.1, *q.v.*). Magnetic fields are commonly found in vast areas such as star-forming regions, with reported intensities ranging from 10^{-10} T to over 4×10^{-4} T.[108] Young stellar objects are known to possess magnetic fields with intensities at the surface of the order of 0.1 to 1 T. These stars are still embedded in the molecular cloud from which they originated and, depending on their spectral types, they may be strong UV sources. This environment is thus favorable for magnetochiral dichroism, combining a strong source of UV radiation and strong magnetic field, along with organic interstellar material. The small anisotropy factors inferred indicate, however, that any *ee* triggered by magnetochiral dichroism under these conditions is likely to be very small ($ee \approx 10^{-6}$).

4.4 Fluid Dynamics: Vortex Motion

Clockwise or anticlockwise stirring has been attempted in numerous experiments aimed at inducing chirality in chemical reactions starting from achiral starting materials. Determined to find the "cosmic force of dissymmetry", and inspired by the motion of the Earth and other celestial bodies in which rotation was in a specific sense rather than the opposite, Pasteur undertook this task, but in vain. In a number of experiments he tried to induce selection in the *dextro-* or *levo-*optical rotation of chemically synthesized molecules by performing reactions in a centrifuge, or even using rotating devices for growing plants, in an attempt to change the sign of optical rotation of their natural products,[109] as outlined in Chapter 3, *The Concept of Chirality*. In general these types of experiments of chiral selection induced by the direction of stirring have usually been disregarded as irreproducible or as artifacts. Indeed, this is the expected outcome of any stirring experiment, since regarded from a simplistic point of view stirring is merely the transmission of angular momentum to the reacting mixture (see Section 3.3.1, *Absence of Chiral Field*). The actual situation is more complex. Vortices are truly chiral objects. In the most simplified view, vortices as chiral inductors can be regarded as the combination of two movements, a rotation-transmitting angular moment ($\vec{L}$, a time-odd axial vector) plus a translation, transmitting linear moment ($\vec{p}$, a time-odd polar vector). This combination of movements mathematically corresponds to the well-known helical topology, to which unambiguous chirality and sense of chirality (helicity) is attributed. The net result is therefore a true chiral effect, analogous to that described in Section 3.3.3, *Experiments under True Chiral Influence*, while in the previous section, 3.3.2, *Experiments under False Chiral Influence*, it was apparent that merely rotation in a gravitational field had only the attributes of false chirality.

Although there is no doubt about the chirality of a vortex, what remains obscure is the force or force fields which originate it. Swirling fluids reveal a concentric gradient of angular velocities around the axis of rotation. This can be attained by including a third pseudo vector in the $\vec{L}\vec{G}$ scenario of Figure 4.11, namely the friction torque, $\vec{\tau}$. In fluid dynamics, the stress tensor describes a complete set of forces involved in all fluid phenomena, including swirling. Its components can be divided into the normal and the shearing stress. The friction torque, $\vec{\tau} = \vec{r} \times \vec{F}$, is a rotational force resulting from the frictional shearing stress ($\vec{F}$), always opposite to the movement of the fluid and therefore opposite to the angular momentum $\vec{L}$ of the bulk liquid, and which creates gradients in its angular velocity. The vector product of the position vector $\vec{r}$ (of an infinitesimal volume of fluid) with respect to the rotating axis ($\vec{r}$ is a time-even polar vector) multiplied by the orthogonal frictional force $\vec{F}$ (also a time-even polar vector) results in a time-even axial vector, $\vec{\tau}$. The time invariance of forces and torques is a consequence of the dependence of these magnitudes upon the second derivative with respect to time (see Section 3.2.2, *The Time Reversion: The Time Reversal Operator, $\hat{T}$*). In real scenarios (*i.e.*, time running forwards), $\vec{L}$ and $\vec{\tau}$ are always antiparallel, so only two chiral states are possible, those of clockwise or anticlockwise rotation with respect to $\vec{G}$, as represented at the top of Figure 4.11.

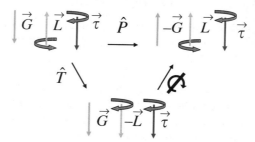

Figure 4.11 The case of a fluid in a spinning vessel. The combined effect of bulk rotation of the fluid (represented by $\vec{L}$) subjected to opposite frictional forces (represented by the torque $\vec{\tau}$) in a gravitational field is apparently a truly chiral influence. It is transformed into its non-superimposable enantiomer by space inversion, $\hat{P}$. The enantiomers are not interconverted by time reversion, $\hat{T}$, as required, although this operation is meaningless since a preferred sense of advance of time, *i.e.*, that of continuous entropic gain, is already implicit in any transport phenomena. In real cases, $\vec{L}$ and $\vec{\tau}$ are always antiparallel, so only two chiral states are possible, those of clockwise or anticlockwise rotation with respect to $\vec{G}$.

Chiral Physical Forces 61

It must be stressed that application of the basic symmetry operations, in particular of $\hat{T}$, to non-equilibrium, not even stationary, systems such as that represented in Figure 4.11, is a loose extension of the principles reported in Section 3.2, devoted entirely to static systems, for which the rate of entropy production is zero, and the two directions of time are equivalent, *i.e.,* entropy is constant. Processes involving transport properties, as is the case of the generation of vortices, occurs with an inevitable rate of production of entropy originated from the resistive forces opposing the motion.[110] From thermodynamic considerations, one direction of time—the real direction of time—is infinitely more probable than the opposite (in which friction seems to "assist" the motion), therefore the two directions of time are nonequivalent. The intrinsically preferred direction of time invalidates, or at least distorts, the significance of the time reversal operation, $\hat{T}$.

In 2001,[111] J. M. Ribó and coworkers reported an experiment in which supramolecular helical structures were formed through a self-assembly process undergoing chiral symmetry breaking.[112] The unexpected result was that the simple effect of clockwise rotary evaporation afforded the necessary chiral influence needed for the stereoselective formation of structures of a given helicity, while reversing the rotation sense afforded the enantiomorphic experiment with generation of structures of reverse helicity.[111,113] The resulting chiral sign of the homoassociate structures was determined by circular dichroism. It was confirmed also that there was solid statistical evidence correlating the sense of rotary evaporation with the sign of the CD spectra obtained from the supramolecular homoassociates. On the other hand, a number of blank unstirred experiments displayed a statistical distribution of helicities without exhibiting any chiral dominance, in what must be taken as a typical chiral symmetry-breaking scenario (see Chapter 6, *Spontaneous Symmetry Breaking*, and Figure 6.2c–d).

At the molecular–supramolecular level these experiments could be interpreted in the following terms. Initially, flat achiral diprotonated porphyrins of zwitterionic structure, such as $H_2TPPS_3^-$ (Figure 4.12, *left*) form homoassociates in aqueous solution, by slow concentration during rotary evaporation. These molecules have a positively charged and hydrogen bonded region inside the porphyrin ring, negatively charged regions in the periphery, and a hydrophobic region around the phenyl substituent. Homoassociation of these disk-shaped charged porphyrins is well defined, and corresponds to a hierarchical process that takes place in different stages.[114] The structure of the first kind of homoassociated aggregate formed is schematically represented in Figure 4.12, *right*. In these homoassociates, zwitterionic porphyrins spontaneously assemble into stacks as a result of intermolecular electrostatic and hydrogen-bonding interactions between the anionic sulfonato groups and the positively charged porphyrin rings. Arrangements of the incoming monomers are possible in line (180°), rotating right (+90°) or rotating left (−90°), with respect to the main strand, given the multiple substitution of the sulfonatophenyl groups in $H_2TPPS_3^-$. This step generates J-shaped aggregates and constitutes the elementary chiral fluctuation, the chiral sense of which can be determined by

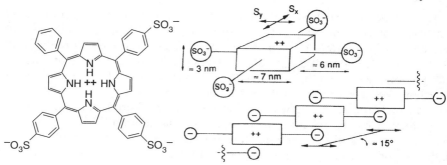

Figure 4.12 *Left*, the *meso*-5, 10, 15-tris(4-sulfonatophenyl)-20-phenyl-porphyrin (H$_2$TPPS$_3^-$); *right*, the form in which tetraphenylsulfonato-porphyrins (H$_2$TPPS$_4^{-2}$) self-assemble into linear 1-D J-aggregate substructures. The next incoming monomeric unit may be attached, rotating either ±90° the direction of the growing strand and generating chiral arrangements of definite helical sense, described as chiral J-aggregates.

circular dichroism (CD). CD provides an appropriate technique for detecting enantiomeric excess in the chiral folding, given the size and chirality of the chromophore, according to the exciton coupling model for π-systems.[115,116] Next, the relative swirling trajectories of the small oligomeric blocks being incorporated are directed by the hydrodynamic forces imparted in the rotating vessel, the overall effect of which is known as vortical stirring. Large H-type bundles are thus assembled from small oligomeric J-aggregates,[117] where preferential asymmetric accretion is imposed by the external vorticity sign, *i.e.*, by its clockwise or anticlockwise rotary motion.

Figure 4.13 represents an actual experiment in which a rotary evaporator slowly triggers the supramolecular aggregation of water-soluble zwitterionic porphyrins such as H$_2$TPPS$_3^-$ (Figure 4.13a) leading to spontaneous chiral symmetry breaking. According to this picture, the helical orientation in the second stage of the assembly process is selected by the sense of the vortex motion (Figure 4.13b). A clockwise or anticlockwise vortex motion provokes, through chiral sign selection, a specific helicity in the process of aggregation, which can be experimentally determined by CD experiments on the supramolecular aggregates. The sign of the CD spectra clearly shows the presence of homoassociates of opposite helicity when prepared under the effect of opposite vortical hydrodynamics. A similar result has been described for a different supramolecular system, involving dendritic structures derived from a zinc porphyrin core.[118] More recently, it was demonstrated that a hydrocarbon solution of these achiral dendritic zinc porphyrins displayed intense CD bands under clockwise stirring, while reversal of stirring resulted in perfect inversion of the sign of the CD spectra, remaining optically inactive at rest.[119] These experiments clearly indicate that a helical arrangement of the J-aggregates was induced by the helical flow. Special care was taken to exclude any contribution of LD (linear dichroism) artifacts in these and all previous experiments which

Chiral Physical Forces

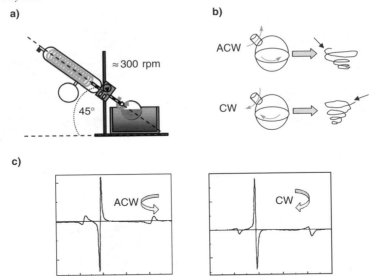

Figure 4.13 (a) Use of a rotary evaporator triggers the supramolecular helical aggregation of $H_2TPPS_3^-$; (b) The sense of rotation generates hydrodynamic forces originating from clockwise (CW) or anticlockwise (ACW) vortices. These intrinsically chiral structures are the actual selectors of the helicity sign displayed by the aggregates, which can be confirmed by CD experiments (c). (Pictures by courtesy of Dr J. M. Ribó.)

relied on CD detection,[120] confirming the existence of an intimate link between macroscopic forces and supramolecular chirality.

In recent experiments it has been actually possible to visualize the structures involved in the above phenomena. Under certain experimental conditions the aggregation and folding processes can be slowed down sufficiently to be tracked by atomic force microscopy (AFM). Evidence for the formation of helical structures under different aggregation regimes, either quiescent or vortex stirring, was obtained by AFM imaging (Figure 4.14).[121] In freshly prepared solutions, the $H_2TPPS_3^-$ porphyrin J-aggregates appear as straight tapes of double layers which begin to fold with time. Comparison of the evolution with time of these J-aggregates in static solutions and vortex-stirred solutions show that chiral-sign selection is caused by the effect of hydrodynamic forces on particle folding. The folding is a spontaneous behavior allegedly driven by entropic factors,[122] and differs dramatically in these two environments. In quiescent solutions the irregular Brownian dynamics result in irregularly folded structures, whereas hydrodynamic gradients (vortex stirring) lead to long-order folding and the formation of helical ribbons; irregular folding and bundles are rarely detected (Figure 4.14).

The aggregation scenario described corresponds to a spontaneous symmetry-breaking process under non-equilibrium conditions, similar to those found in the crystallization of conglomerates (see Section 6.2, *Spontaneous Symmetry Breaking in Crystallization*). In such non-equilibrium situations, symmetry breaking can

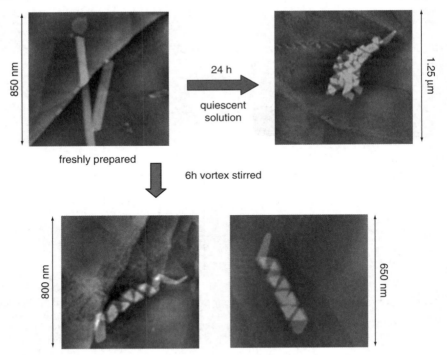

Figure 4.14 Comparison of the shape evolution of J-aggregates of the porphyrin $H_2TPPS_3^-$ in stagnant and vortex-stirred solutions. The chiral-sign selection, or helicity, of the aggregates is due to the action of hydrodynamic gradients (vortex motion) on particle folding. Static solutions result in irregularly folded aggregates, whereas vortex stirring leads to long-order folding and the formation of helical ribbons. (Pictures from Reference 121, with permission.)

occur as a consequence of weak stochastic fluctuations coupled with an autocatalytic process. In the absence of any external chiral influence, the chirality sign of the supramolecular structures formed during the aggregation process is usually determined by small random fluctuations, and the symmetry is only restored over a large number of experiments, as in the crystallization of conglomerates. This perfectly symmetrical bifurcation situation is, however, modified in this case by a macroscopic physical force. In this experiment, it was statistically confirmed that the direction of the stirring vortex, a weak external physical force, selects the chirality sign of these aggregates, introducing a bias in the otherwise random selection due to stochastic fluctuations. The only two other examples—with experimental validation—of a physical force chirally perturbing the composition of an achiral chemical sample have been concerned with interaction with light.

It is accepted that aggregation and self-assembly played an important role in prebiotic chemistry.[123] Simple achiral molecules presumably evolved, increasing in size and complexity, in many cases through processes mediated by supramolecular interactions. Along with this evolution, an increase in the number of functions capable of being carried out ultimately brought about the appearance of elemental forms endowed with the attributes of living organisms. The experiments of vortex chiral sign selection provide a remarkable example of how chiral information can be transmitted between the macroscopic and molecular levels. At the macroscopic level, the occurrence of vortex phenomena is common in nature. In the atmospheric sciences, vorticity is a property characterizing large-scale rotation of air masses. On Earth, tropical cyclones, also referred to by various names depending on their location and strength such as hurricane, typhoon, tropical storm and cyclonic storm, are atmospheric manifestations of vortices. Similar, but much greater, vortices are also seen on other planets, such as the permanent Great Red Spot on Jupiter and the intermittent Great Dark Spot on Neptune. In physical oceanography, more persistent and involving a denser, more viscous fluid, the ocean currents are generated by the forces acting upon the water, including the Earth's rotation, together with other factors such as temperature and wind. The Earth's rotation imparts an acceleration to any mass or fluid moving relative to it, the Coriolis effect. Having an opposite sign in each hemisphere, the Coriolis effect is responsible for the initial cyclonic rotation and also has a profound influence on the flow of the oceans. The obvious weak point of this scenario of chiral transfer is that the Earth's two hemispheres should produce opposing effects, thereby generating enantiomers of the prebiotic molecular assemblies, but this is not necessarily the case. The anomaly is easily explained if it is assumed that early terrestrial life developed in a local spot in one hemisphere, *i.e.*, on or near exposed solid ground. Indeed, the asymmetry of the continental distribution between hemispheres might prematurely point toward a deterministic imbalance. Extrapolation of the continental drift beyond 3800 million years ago during prebiotic evolution is at most very uncertain. It was a period of intense tectonic activity. Information on paleolatitude has been provided from paleomagnetic data,[124] from the Pilbara (current Western Australia) and Kaapvaal (current South Africa) cratons, to test the hypothesis that these cratons were joined as part of an archaean supercontinent, Vaalbara,[125] proposed as being the earliest supercontinent between 3300 and 3600 million years ago. Unfortunately this does not exclude other possible locations now submerged and unstudied.

Whether these chiral hydro- or aerodynamic forces might have conferred a chiral sign to the evolving prebiotic molecules through their supramolecular interactions remains unknown, and requires more study. However, what has been proved at the laboratory level is that selection of the molecular chirality sign is possible by application of macroscopic chiral forces, in particular by a stirring vortex.

References

1. "Standard Model" from *Encyclopædia Britannica* Online: http://search.eb.com/eb/article?eu = 71193.
2. M. L. Goldhaber, L. Grodzins and A. Sunyar, *Phys. Rev.*, 1958, **109**, 1015–1017.

3. E. P. Wigner, *Z. Phys.*, 1927, **43**, 624–652.
4. T. D. Lee and C. N. Yang, *Phys. Rev.*, 1956, **105**, 1413–1415.
5. C. S. Wu, E. Ambler, R. W. Hayward, D. D. Hoppes and R. P. Hudson, *Phys. Rev.*, 1957, **105**, 1413–1415.
6. R. G. Sachs, *The Physics of Time Reversal*, University of Chicago Press, Chicago, 1987, pp. 188–233.
7. A. Das and T. Ferbel, *Introduction to Nuclear and Particle Physics*, World Scientific Publishing Co., River Edge, NJ, 2003, pp. 287–312.
8. See Reference 6, p. 160.
9. S. L. Glashow, *Rev. Mod. Phys.*, 1980, **52**, 539–543.
10. S. Weinberg, *Rev. Mod. Phys.*, 1980, **52**, 515–523.
11. A. Salam, *Rev. Mod. Phys.*, 1980, **52**, 525–538.
12. S. Weinberg, *Phys. Rev. Lett.*, 1967, **19**, 1264–1266.
13. B. Zel'dovich, *Sov. Phys. JEPT*, 1959, **9**, 682.
14. (a) Y. Yamagata, *J. Theoret. Biol.*, 1966, **11**, 495–498; (b) D. Rein, *J. Mol. Evol.*, 1974, **4**, 15–22.
15. (a) L. D. Barron, *Mol. Phys.,* 1981, **43**, 1395–1406; (b) L. D. Barron, *Chem. Soc. Rev.*, 1986, **15**, 189–223.
16. UA2 Collaboration, P. Bagnaia, *et al.*, *Phys. Lett. B*, 1983, **129**, 130–140.
17. (a) B. Frois and M. A. Bouchiat, *Parity Violation in Atoms and Polarized Electron Scattering*, World Science, Singapore, 1999; (b) M. A. Bouchiat and L. Pottier, *Science*, 1986, **234**, 1203–1210; (c) J. Gueña, M. Lintz and M. A. Bouchiat, *Modern Phys. Lett. A*, 2005, **20**, 375–389.
18. (a) A. Derevianko, *Phys. Rev. Lett.*, 2000, **85**, 1618–1621; (b) A. Derevianko, *Phys. Rev. A*, 2001, **65**, 012106/1; (c) A. I. Milstein, O. P. Sushkov and I. S. Terekhov, *Phys. Rev. Lett.*, 2002, **89**, 283003/4.
19. (a) L. M. Barkov and M. S. Zolotarev, *Phys. Lett. B*, 1979, **85B**, 308–313; (b) J. H. Hollister, G. R. Apperson, L. L. Lewis, T. P. Emmons, T. G. Vold and E. N. Fortson, *Phys. Rev. Lett.*, 1981, **46**, 643–646.
20. T. P. Emmons, J. M. Reeves and E. N. Fortson, *Phys. Rev. Lett.*, 1983, **51**, 2089–2092.
21. (a) P. S. Drell and E. D. Commins, *Phys. Rev. Lett.*, 1984, **53**, 968–971; (b) P. A. Vetter, D. M. Meekhof, P. K. Majumder, S. K. Lamoreaux and E. N. Fortson, *Phys. Rev. Lett.*, 1995, **74**, 2658–2661.
22. (a) S. L. Gilbert, M. C. Noecker, R. N. Watts and C. E. Wieman, *Phys. Rev. Lett.*, 1985, **55**, 2680–2683; (b) C. E. Wieman, M. C. Noecker, B. P. Masterson and J. Cooper, *Phys. Rev. Lett.*, 1987, **58**, 1738–1741.
23. (a) D. M. Meekhof, P. Vetter, P. K. Majumder, S. K. Lamoreaux and E. N. Fortson, *Phys. Rev. Lett.*, 1993, **71**, 3442–3445; (b) D. M. Meekhof, P. A. Vetter, P. K. Majumder, S. K. Lamoreaux and E. N. Fortson, *Physical Review A*, 1995, **52**, 1895–1908.
24. M. A. Bouchiat and C. C. Bouchiat, *Phys. Lett. B*, 1974, **48**, 111–114.
25. C. Y. Prescott, W. B. Atwood, R. L. A. Cottrell, H. DeStaebler, E. L. Garwin, A. Gonidec, R. H. Miller, L. S. Rochester, T. Sato, D. J. Sherden, C. K. Sinclair, S. Stein, R. E. Taylor, J. E. Clendenin,

V. W. Hughes, N. Sasao, K. P. Schüler, M. G. Borghini, K. Lübelsmeyer and W. Jentschke, *Phys. Lett. B*, 1978, **77**, 347–352.
26. W.-Y. P. Hwang, E. M. Henley and G. A. Miller, *Ann. Phys.*, 1981, **137**, 378–440.
27. (a) C. S. Wood, S. C. Bennett, D. Cho, B. P. Masterson, J. L. Roberts, C. E. Tanner and C. E. Wieman, *Science*, 1997, **275**, 1759–1763; (b) J. Guéna, D. Chauvat, P. Jacquier, E. Jahier, M. Lintz, S. Sanguinetti, A. Wasan, M. A. Bouchiat, A. V. Papoyan and D. Sarkisyan, *Phys. Rev. Lett.*, 2003, **90**, 143001/4.
28. M. Gaillard, P. Grannis and F. Sciulli, *Rev. Mod. Phys.*, 1999, **71**, S96–S111.
29. A. Bakasov, T.-K. Ha and M. Quack, *J. Chem. Phys.*, 1998, **109**, 7263–7285.
30. R. A. Hegstrom, D. W. Rein and P. G. H. Sandars, *J. Chem. Phys.*, 1980, **73**, 2329–2341.
31. (a) A. L. Barra and J. B. Robert, *Mol. Phys.*, 1996, **88**, 875–886; (b) A. L. Barra, J. B. Robert and L. Wiesenfeld, *Europhys. Lett.*, 1988, **5**, 217–222.
32. (a) C. C. Bouchiat and M. A. Bouchiat, *J. Phys.*, 1974, **35**, 899–927; (b) C. C. Bouchiat and M. A. Bouchiat, *J. Phys.*, 1975, **36**, 493–509.
33. (a) R. Berguer and M. Quack, *J. Chem. Phys.*, 2000, **112**, 3148–3158; (b) G. Laubender and R. Berguer, *ChemPhysChem.*, 2003, **4**, 395–399.
34. T. L. V. Ulbricht and F. Vester, *Tetrahedron*, 1962, **18**, 629–637.
35. (a) D. Rein, R. Hegström and P. Sandars, *Phys. Lett. A*, 1979, **71**, 499–502; (b) R. Hegström, D. Rein and P. Sandars, *J. Chem. Phys.*, 1980, **73**, 2329–2341.
36. (a) A. J. MacDermott, in *Physical Origin of Homochirality in Life*, ed. D. B. Cline, American Institute of Physics, Woodbury, NY, 1996, p. 241; (b) G. E. Tranter and A. McDermott, *Chem. Phys. Lett.*, 1986, **130**, 120–122; (c) G. E. Tranter, *Mol. Phys.*, 1985, **56**, 825–838; (d) S. F. Mason and G. E. Tranter, *J. Chem. Soc., Chem. Commun.*, 1983, 117–119; (e) S. F. Mason and G. E. Tranter, *Mol. Phys.*, 1984, **53**, 1091–1111; (f) A. McDermott and G. E. Tranter, *Chem. Phys. Lett.*, 1989, **163**, 1–4; (g) G. E. Tranter, *J. Theor. Biol.*, 1986, **119**, 467–479.
37. A. Macdermott, *J. Orig. Life Evol. Biosphere*, 1995, **25**, 191–199.
38. P. Cintas, *ChemPhysChem*, 2001, **2**, 409–410.
39. W. Bonner, *Chirality*, 2000, **12**, 114–126.
40. M. Quack, *Angew. Chem. Int. Ed.*, 2002, **41**, 4618–4630.
41. A. Bakasov and M. Quack, *Chem. Phys. Lett.*, 1999, **303**, 547–557.
42. (a) P. Lazzeretti and R. Zanasi, *Chem. Phys. Lett.*, 1997, **279**, 349–354; (b) F. Faglioni and P. Lazzeretti, *Phys. Rev. E*, 2002, **65**, 011904/11.
43. (a) J. K. Laerdahl, P. Schwerdtfeger and H. M. Quiney, *Phys. Rev. Lett.*, 2000, **84**, 3811–3814; (b) J. K. Laerdahl and P. Schwerdtfeger, *Phys. Rev. A*, 1999, **60**, 4439–4453.
44. R. Zanasi and P. Lazzeretti, *Chem. Phys. Lett.*, 1998, **286**, 240–242.
45. (a) R. Berguer and M. Quack, *ChemPhysChem*, 2000, **1**, 57–60; (b) J. K. Laerdahl, R. Wesendrup and P. Schwerdtfeger, *ChemPhysChem*, 2000, **1**, 60–62.

46. T. Watanabe, K. Morihashi, O. Takahashi, T. Kitayama, T. Yagi and O. Kikuchi, *J. Mol. Struct.*, 2004, **671**, 119–123.
47. R. Zanasi, P. Lazzeretti, A. Ligabue and A. Soncini, *Phys. Rev. E*, 1999, **59**, 3382–3385.
48. M. Quack and J. Stohner, *Chimia*, 2005, **59**, 530–538.
49. (a) H. Buschmann, R. Thede and D. Heller, *Angew. Chem. Int. Ed.*, 2000, **39**, 4033–4036; (b) M. Quack, *Chimia*, 2003, **57**, 147–160.
50. V. S. Letokhov, *Phys. Lett.*, 1975, **53A**, 275–277.
51. (a) M. Quack and J. Stohner, *J. Chem. Phys.*, 2003, **119**, 11228–11240; (b) A. Bauder, A. Beil, D. Luckhaus, F. Muller and M. Quack, *J. Chem. Phys.*, 1997, **106**, 7558–7570.
52. (a) O. N. Kompanets, A. R. Kukudzhanov, V. S. Letokhov and L. L. Gervits, *Opt. Commun.*, 1976, **19**, 414–416; (b) C. Daussy, T. Marrel, A. Amy-Klein, C. T. Nguyen, C. J. Bordé and C. Chardonnet, *Phys. Rev. Lett.*, 1999, **83**, 1554–1557.
53. (a) G. Laubender and R. Berger, *ChemPhysChem.*, 2003, **4**, 395–399; (b) J. B. Robert and A. L. Barra, *Chirality*, 2001, **13**, 699–702. (c) R. Bast, P. Schwerdtfeger and T. Saue, *J. Chem. Phys.* 2006, **125**, 064504/7.
54. L. N. Ivanov and V. S. Letokhov, *J. Chem. Phys.*, 1997, **106**, 6045–6050.
55. A. V. Baluev, L. V. Inzhechik, E. V. Melnikov, B. I. Rogozev, A. S. Khlebnikov, V. G. Tsinoev and V. M. Cherepanov, *JEPT Lett.*, 1986, **43**, 656–658.
56. R. A. Harris, *Chem. Phys. Lett.*, 2002, **365**, 343–346.
57. (a) W-Q. Wang, F. Yi, Y. Ni, Z. Zhao, X. Jin, Y. Tang, *J. Biol. Phys.*, 2000, **26**, 51–65; (b) W-Q. Wang, W. Min, Z. Liang, L-Y. Wang, L. Chen and F. Deng, *Biophys.Chem.*, 2003, **103**, 289–298.
58. A. Szabó-Nagy and L. Keszthelyi, *Proc. Natl. Acad. Sci. USA*, 1999, **96**, 4252–4255.
59. M. Lahav, I. Weissbuch, E. Shavit, C. Reiner, G. J. Nicholson and V. Schurig, *Orig. Life Evol. Biosphere*, 2006, **36**, 151–170.
60. (a) T. Marrel, M. Ziskind, C. Daussy and C. Chardonnet, *J. Mol. Struct.*, 2001, **599**, 195–209; (b) M. Quack and J. Stohner, *J. Chem. Phys.*, 2003, **119**, 11228–11240.
61. J. Crassous, F. Monier, J.-P. Dutasta, M. Ziskind, C. Daussy, C. Grain and C. Chardonnet, *ChemPhysChem.*, 2003, **4**, 541–548.
62. M. Quack and J. Stohner, *Chirality*, 2001, **13**, 745–753.
63. (a) P. Schwerdtfeger and R. Bast, *J. Am. Chem. Soc.*, 2004, **126**, 1652–1653; (b) P. Schwerdtfeger, J. Gierlich and T. Bollwein, *Angew. Chem. Int. Ed.*, 2003, **42**, 1293–1296.
64. L. D. Barron, *Chem. Soc. Rev.*, 1986, **15**, 189–223.
65. F. Vester, T. L. V. Ulbricht and H. Krauss, *Naturwissenschaften*, 1959, **46**, 68–69.
66. T. L. V. Ulbricht and F. Vester, *Tetrahedron*, 1962, **18**, 629–637.
67. W. A. Bonner, *Chirality*, 2000, **12**, 114–126.
68. D. L. Anderson, in *Theory of the Earth*, 1999, available on the web at http://resolver.caltech.edu/CaltechBOOK:1989.001.

69. A. G. Griesbeck and U. Meierhenrich, *Angew. Chem. Int. Ed.*, 2002, **41**, 3147–3154.
70. W. A. Bonner, *Top. Stereochem.*, 1988, **18**, 1–96.
71. J. J. Flores, W. A. Bonner and G. A. Massey, *J. Am. Chem. Soc.*, 1977, **99**, 3622–3625.
72. H. Nishino, A. Kosaka, G. A. Hembury, H. Shitomi, H. Onuki and Y. Inoue, *Org. Lett.*, 2001, **3**, 921–924.
73. G. Prumper, J. Viefhaus, S. Cvejanovic, D. Rolles, O. Gessner, T. Lischke, R. Hentges, C. Wienberg, W. Mahler, U. Becker, B. Langer, T. Prosperi, N. Zema, S. Turchini, B. Zada and F. Senf, *Phys. Rev. A*, 2004, **69**, 062717/7.
74. U. J. Meierhenrich, L. Nahon, C. Alcaraz, J. H. Bredehöft, S. V. Hoffmann, B. Barbier and A. Brack, *Angew. Chem. Int. Ed.*, 2005, **44**, 5630–5634.
75. (a) A. Moradpour, J. F. Nicoud, G. Balavoine, H. Kagan and G. Tsoucaris, *J. Am. Chem. Soc.*, 1971, **93**, 2353–2354; (b) H. Kagan, A. Moradpour, J. F. Nicoud, G. Balavoine, R. H. Martin and J. P. Cosyn, *Tetrahedron Lett.*, 1971, **27**, 2479–2482.
76. (a) W. J. Bernstein, M. Calvin and O. Buchardt, *J. Am. Chem. Soc.*, 1972, **94**, 494–497; (b) W. J. Bernstein, M. Calvin and O. Buchardt, *Tetrahedron Lett.*, 1972, **13**, 2195–2198; (c) W. J. Bernstein, M. Calvin and O. Buchardt, *J. Am. Chem. Soc.*, 1973, **95**, 527–532.
77. U. J. Meierhenrich and W. H-P. Thiemann, *Orig. Life Evol. Biosphere*, 2004, **34**, 111–121.
78. R. D. Wolstencroft, *Astronomical Sources of Circularly Polarized Light and their Role in Determining Molecular Chirality on Earth*, IAU Symp., **112**, The Search for Extraterrestrial Life, ed. M. D. Papagiannis, D. Reidel Publishing, Dordrecht, 1985, 171–175.
79. G. Goodman and M. E. Gershwin, *Experimental Biol. Med.*, 2006, **231**, 1587–1592.
80. (a) W. A. Bonner, *Origins Life Evol. Biosph.*, 1991, **21**, 59–111; (b) L. Keszthelyi, *Quart. Rev. Biophys.*, 1995, **28**, 473–507.
81. L. Mörtberg, *Nature*, 1971, **232**, 105–107.
82. A. Jorissen and C. Cerf, *Orig. Life Evol. Biosph.*, 2002, **32**, 129–142.
83. E. L. Gibb, D. C. Whittet and J. E. Chiar, *Astrophys. J.*, 2001, **558**, 702–716.
84. M. P. Bernstein, S. A. Sandford, L. J. Allamandola, S. Chang and M. A. Scharberg, *Astrophys. J.*, 1995, **454**, 327–344.
85. G. M. Muñoz Caro, U. J. Meierhenrich, W. A. Schutte, B. Barbier, A. Arcones Segovia, H. Rosenbauer, W. H-P. Thiemann, A. Brack and J. M. Greenberg, *Nature*, 2002, **416**, 403–406.
86. M. P. Bernstein, J. P. Dworkin, S. A. Sandford, G. W. Cooper and L. J. Allamandola, *Nature*, 2002, **416**, 401–403.
87. (a) M. Nuevo, U. J. Meierhenrich, G. M. Muñoz Caro, E. Dartois, L. d'Hendecourt, D. Deboffle, G. Auger, D. Blanot, J. H. Bredehöft and L. Nahon, *Astron. Astrophys.*, 2006, **457**, 741–751; (b) M. Nuevo, U. J.

Meierhenrich, L. d'Hendecourt, G. M. Muñoz Caro, E. Dartois, D. Deboffle, W. H. P. Thiemann, J. H. Bredehöft and L. Nahon, *Adv. Space Res.*, 2007, **39**, 400–404.
88. J. Bailey, A. Chrysostomou, J. H. Hough, T. M. Gledhill, A. McCall, S. Clark, F. Ménard and M. Tamura, *Science*, 1998, **281**, 672–674.
89. (a) J. Bailey, *Orig. Life Evol. Biosphere*, 2001, **31**, 167–183; (b) F. Menard, A. Chrysostomou, T. Gledhill, J. H. Hough and J. Bailey, *Astronl. Soc. Pacific Conf. Ser.*, 2000, **213**, 349–354.
90. W. A. Bonner and E. Rubenstein, *BioSystem*, 1987, **20**, 99–111.
91. W. A. Bonner, *Orig. Life Evol. Biosphere*, 1991, **21**, 59–111.
92. W. A. Bonner, E. Rubenstein and G. S. Brown, *Orig. Life Evol. Biosphere*, 1999, **29**, 329–332.
93. S. F. Mason, *Nature*, 1997, **389**, 804.
94. J. Oró, *Nature*, 1961, **190**, 389–390.
95. J. Podlech, *Angew. Chem. Int. Ed.*, 1999, **38**, 477–478.
96. B. A. Cohen and C. F. Chyba, *Icarus*, 2000, **145**, 272–281.
97. J. L. Bada and G. D. McDonald, *Icarus*, 1995, **114**, 139–143.
98. S. Pizzarello, *Orig. Life Evol. Biosphere*, 2004, **34**, 25–34.
99. (a) L. J. Allamandola, A. G. G. M. Tielens and J. R. Barker, *Astrophys. J.*, 1985, **290**, L25–L28; (b) D. M. Hudgins, *Polycyclic Aromat. Compd.*, 2002, **22**, 469–488; (c) L. J. Allamandola, S. A. Sandford and B. Wopenka, *Science*, 1987, **237**, 56–59.
100. M. P. Bernstein, S. A. Sandford, L. J. Allamandola, J. S. Gillette, S. J. Clemett and R. N. Zare, *Science*, 1999, **283**, 1135–1138.
101. (a) M. P. Groenewege, *Molec. Phys.*, 1962, **5**, 541–563; (b) N. B. Baranova, Y. V. Bogdanov and B. Y. Zel'dovich, *Optics Comm.*, 1977, **22**, 243–247.
102. (a) N. B. Baranova and B. Ya Zel'dovich, *Mol. Phys.*, 1979, **38**, 1085–1098; (b) G. Wagnière and A. Meier, *Experientia*, 1983, **39**, 1090–1091.
103. L. D. Barron and J. Vrbancich, *Mol. Phys.*, 1984, **51**, 715–730.
104. G. L. J. A. Rikken and E. Raupach, *Nature*, 1997, **390**, 493–494.
105. G. L. J. A. Rikken and E. Raupach, *Nature*, 2000, **405**, 932–935.
106. P. Kleindienst and G. H. Wagnière, *Chem. Phys. Lett.*, 1998, **288**, 89–97.
107. L. D. Barron, *Nature*, 2000, **405**, 895–896.
108. (a) R. Genzel and J. Stutzki, *Ann. Rev. Astron. Astrophys.*, 1989, **27**, 41–86; (b) C. Heiles, A. A. Goodman, C. F. McKee and E. G. Zweibel, *Magnetic Fields in Star-Forming Regions: Observations*,in *Protostars and Planets III*, ed. E. H. Levy and J. I. Lunine, The University of Arizona Press, Tucson/London, 1993, 279–326.
109. (a) L. Pasteur, *Bull. Soc. Chim. Fr.*, 1884, **41**, 215–221; (b) L. Pasteur Valerie-Radot, Oeuvres de Pasteur, Masson et Cie., Paris, 1922, vol. I, 61–64, 369–386.
110. R. R. Birss, *Symmetry and Magnetism*, North-Holland, Amsterdam, 1966, 2nd Ed, 99–101.
111. J. M. Ribo, J. Crusats, F. Sagues, J. Claret and R. Rubires, *Science*, 2001, **292**, 2063–2066.

112. O. Ohno, Y. Kaizu and H. Kobayashi, *J. Chem. Phys.*, 1993, **99**, 4128–4139.
113. R. Rubires, J.-A. Farrera and J. M. Ribó, *Chem. Eur. J.*, 2001, **7**, 436–446.
114. (a) R. F. Pasternack, P. R. Huber, P. Boyd, G. Engasser, L. Francesconi, E. Gibbs, P. Fasella, G. C. Venturo and L. de C. Hinds, *J. Am. Chem. Soc.*, 1972, **94**, 4511–4517; (b) T. P. G. Sutter, R. Rahimi, P. Hambright, J. C. Bommer, M. Kumar and P. Neta, *J. Chem. Soc. Faraday Trans.*, 1993, **89**, 495–502; (c) J. M. Ribó, J. Crusats, J-A. Farrera and M. L. Valero, *J. Chem. Soc. Chem. Commun.*, 1994, 681–682; (d) R. Rubires, J. Crusats, Z. El-Hachemi, T. Jaramillo, M. López, E. Valls, J-A. Farrera, J. M. Ribó, *New J. Chem.*, 1999, 189–198; (e) C. Escudero, Z. El-Hachemi, J. Crusats and J. M. Ribó, *J. Porphyrins Phthalocyanines*, 2006, **9**, 852–863; (f) N. Micali, V. Villari, M. A. Castriciano, A. Romeo and L. Monsu Scolaro, *J. Phys. Chem. B*, 2006, **110**, 8289–8295.
115. N. Harada and K. Nakanishi, *Circular Dichroism Spectroscopy*, University Science Books, Mill Valley, CA, 1983.
116. J. M. Ribó, J. M. Bofill, J. Crusats and R. Rubires, *Chem. Eur. J.*, 2001, **7**, 2733–2737.
117. J. Crusats, J. Claret, I. Diez-Perez, Z. El-Hachemi, H. Garcia-Ortega, R. Rubires, F. Sagues and J. M. Ribó, *Chem. Commun.*, 2003, 1588–1589.
118. T. Yamaguchi, T. Kimura, H. Matsuda and T. Aida, *Angew. Chem. Int. Ed.*, 2004, **43**, 6350–6355.
119. A. Tsuda, M. A. Alam, T. Harada, T. Yamaguchi, N. Ishii and T. Aida, *Angew. Chem. Int. Ed.*, 2007, **46**, 8198–8202.
120. M. Wolffs, S. J. George, Z. Tomovic, S. C. J. Meskers, A. P. H. J. Schenning and E. W. Meijer, *Angew. Chem. Int. Ed.*, 2007, **46**, 8203–8205.
121. C. Escudero, J. Crusat, I. Diez-Perez, Z. El-Hachemi and J. M. Ribó, *Angew. Chem. Int. Ed.*, 2006, **45**, 8032–8035.
122. Y. Snir and R. D. Kamien, *Science*, 2005, **35**, 1067.
123. (a) D. Deamer, S. Singaram, S. Rajamani, V. Kompanichenko and S. Guggenheim, *Phil. Trans. R. Soc. B.*, 2006, **361**, 1809–1818; (b) A. W. Schwartz, *Curr. Biol.*, 1997, **7**, R477–R499; (c) D. Deamer, S. I. Kuzina, A. I. Mikhailov, E. I. Maslikova and S. A. Seleznev, *J. Evol. Biochem. Physiol.*, 1991, **27**, 212–217.
124. M. T. D. Wingate, *South African J. Geol.*, 1998, **101**, 257–274.
125. (a) T. E. Zegers, M. J. De Wit, J. Dann and S. H. White, *Terra Nova*, 1998, **10**, 250–259; (b) D. R. Nelson, A. F. Trendall and W. Altermann, *Precambrian Res.*, 1999, **97**, 165–189.

CHAPTER 5
Mechanisms of Amplification

5.1 Background

The remarkably small magnitude of the energies involved in the calculations mentioned in Chapter 4 (*e.g.*, Section 4.2.5, *Quantification of the Parity Violation in Molecules:* ΔE_{pv}) means that only an exceedingly small excess of one favored enantiomer may be produced over the other. Conversely, at the biomolecular level the preference for one enantiomer is complete. If a connection between these two phenomena is to be proposed, some kind of highly efficient, or perhaps slow but relentless, mechanism of amplification should operate before any perceptible enantiomeric excess is generated. Similar arguments apply to the amplification of the chirality of a cryptochiral mixture (Section 2.2.1, *Amplification of Tiny Stochastic Imbalances*), such as the standard deviation from equimolecular in the number of enantiomers in a racemic mixture.

5.2 Autocatalysis

5.2.1 The Frank Model

Dating back to 1953,[1] F. C. Frank developed a simple scheme of reactions which is still widely accepted at the present time, and is illustrated in Scheme 5.1.[2] This model proposes the amplification of a small initial asymmetry in chiral autocatalytic reactions, which is the key feature of the process. An open flux reactor is fed with a flow of achiral reagents, A and B, which react to give products L and D. The process is autocatalytic, in other words the products L and D catalyze their own formation. Due to symmetry, the rate constants for both the initial and autocatalytic reactions, (1) and (2), and (3) and (4), respectively, must be equal. L and D are therefore produced with equal preference. Finally, we assume that the two enantiomers react with one another and are converted to an

---

The Origin of Chirality in the Molecules of Life
Albert Guijarro and Miguel Yus
© Albert Guijarro and Miguel Yus, 2009
Published by the Royal Society of Chemistry, www.rsc.org

Mechanisms of Amplification

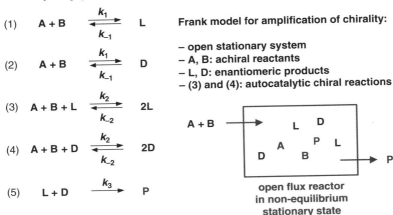

(1) $A + B \underset{k_{-1}}{\overset{k_1}{\rightleftharpoons}} L$

(2) $A + B \underset{k_{-1}}{\overset{k_1}{\rightleftharpoons}} D$

(3) $A + B + L \underset{k_{-2}}{\overset{k_2}{\rightleftharpoons}} 2L$

(4) $A + B + D \underset{k_{-2}}{\overset{k_2}{\rightleftharpoons}} 2D$

(5) $L + D \overset{k_3}{\rightarrow} P$

Frank model for amplification of chirality:
- open stationary system
- A, B: achiral reactants
- L, D: enantiomeric products
- (3) and (4): autocatalytic chiral reactions

open flux reactor in non-equilibrium stationary state

Scheme 5.1 The Frank model for amplification of chirality. An open non-equilibrium system with an input of achiral reagents, A and B, initially generates a racemic mixture, L + D, through direct (1 and 2) and autocatalyzed reactions (3 and 4). At the same time the inactive product, P, is pumped out to permit the system become stationary (5).

inactive compound P, which is pumped out from the reactor (5), in what Frank described as "mutual antagonism". This final step is necessary for chiral amplification,[3] and the removal of P allows a stationary state to be achieved.

By pumping reagents in and products out, a stationary state is reached. The concentration of A and B remain constant inside the reactor; their concentrations will be used as a variable to explore the system at different concentrations. The kinetic equations describing an open flux reactor with these characteristics are given in Figure 5.1, Equations (6) and (7); these can be deduced from Reactions (1) to (5), above, by including the kinetic balances for L and D. To enable the solution of these kinetic equations it is convenient to define the following variables: $\lambda = [A][B]$; $\alpha = ([L]-[D])/2$; and $\beta = ([L]+[D])/2$. When Equations (6) and (7) are rewritten in terms of α and β and calculated to find the stationary states, i.e., $d\alpha/dt = 0$ and $d\beta/dt = 0$, we observe two possible general scenarios. If the concentration of A and B remain below a critical point, the system is entirely racemic, Scenario (a). On the other hand, as the input concentration of A + B is increased, the racemic process becomes metastable and switches spontaneously into one of the two possible branches, giving either enrichment of L or enrichment of D, breaking the racemic chiral symmetry, Scenario (b).

The graphic representation of these scenarios in Figure 5.1 shows the evolution of α as a function of λ. α is proportional to the optical rotation of the reaction mixture, and is zero unless the mixture becomes nonracemic; λ increases with the concentration of the achiral input. The two possible scenarios are indicated in red: Scenario (a) operates at low values of λ (low concentrations of A + B), and the system is racemic ($\alpha = 0$) up to a critical value, λ_C.

Kinetic Equations of the Frank's model

(6) $\dfrac{d[L]}{d[t]} = k_1[A][B] - k_{-1}[L] + k_2[A][B][L] - k_{-2}[L]^2 - k_3[L][D]$

(7) $\dfrac{d[D]}{d[t]} = k_1[A][B] - k_{-1}[D] + k_2[A][B][D] - k_{-2}[D]^2 - k_3[L][D]$

change of variables

$\lambda = [A][B] \qquad \alpha = \dfrac{[L]-[D]}{2} \qquad \beta = \dfrac{[L]+[D]}{2}$

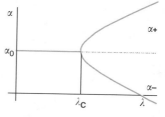

graphic representation of solutions

- Scenario (a): $\lambda < \lambda_C$, the system is racemic, $\alpha_0 = 0$
- Scenario (b): $\lambda > \lambda_C$, the system becomes metastable and leads to a asymmetric dissipative structure with two possible outcomes, $\alpha+$ and $\alpha-$.

Figure 5.1 Kinetic equations of the Frank's model, corresponding to Reactions (1) to (5) in Scheme 5.1. After a change of variables (λ, α and β) the search of stationary states produces the graph on the right. Two different scenarios are found: at low concentration of achiral supply (A and B) the system is racemic, whereas at higher concentrations beyond the critical point (λ_C) the system becomes metastable and tends spontaneously to develop homochirality, with equal probability toward each branch ($\alpha+$ or $\alpha-$).

Beyond λ_C, Scenario (b) operates, in which the system is nonracemic and $\alpha \neq 0$. Two equally probable solutions of the kinetic equations are possible. At concentrations of reagents higher than λ_C, the racemic system becomes metastable (dotted blue line) and hypersensitive to small chiral perturbations. The chiral perturbation may be external to the system, or part of it, and the process that is triggered has been denominated chiral sign selection. For example, a small excess of the parity-violating favored enantiomer in a racemic mixture, or random fluctuations, which are several orders of magnitude larger than ΔE_{pv} (compare Section 4.2.5, *Quantification of the Parity Violation in Molecules: ΔE_{pv}*, and Section 2.2.1, *Amplification of Tiny Stochastic Imbalances*) represent examples of internal chiral perturbations. In the absence of chiral perturbations, random fluctuations will determine the particular homochiral channel adopted by the system, 50% of the time statistically favoring the L- and 50% the D-enantiomer. This is known as a "dissipative structure", generated by dissipative forces (random molecular movement) and which could generate asymmetry in systems far from equilibrium.[2]

With the exception of Soai's autocatalyzed reaction (Section 5.2.4, *Asymmetric Autocatalytic Reactions*), no chemical reaction has so far produced chiral asymmetry in this simple manner. Symmetry breaking does however occur during crystallization of certain compounds from supersaturated liquids. Sodium chlorate, for example, and also 1,1'-binaphthyl, which are optically inactive in solution or in the molten phase, respectively, can produce optically active crystallites in this manner (Section 6.2, *Spontaneous Symmetry Breaking in Crystallization*).

Mechanisms of Amplification

Another scenario of autocatalytic symmetry breaking is the formation of chiral mesophases in aggregation processes forming chiral liquid crystals.[4,5] The self-assembly of these supramolecular structures spontaneously produces chiral symmetry breaking, in which helical colloidal particles are generated in a nonracemic manner from achiral monomeric units. In the absence of an external chiral influence this gives a random unpredictable outcome in chiral sign (Section 4.4, *Fluid Dynamics: Vortex Motion*).

5.2.2 Theoretical Models Derived from Frank's Original Model[6]

The original model described by Frank is the simplest example of an autocatalytic process which evolves towards homochirality by amplification of a tiny initial chiral imbalance. The necessary elements are a direct autocatalytic reaction of L and D, and their mutual elimination. In variations of this model, there may be two incoming achiral compounds (bimolecular reaction of A + B → L + D, Section 5.2.1, *The Frank Model*),[7] only one (unimolecular reaction: A → L + D),[8] a low-barrier racemizing compound.[9] the Frank model itself, reactions take place within an open-flow reactor which allows a stationary state to be maintained under conditions far from thermodynamic equilibrium.

Such an open-flow system is necessary to maintain a non-equilibrium regime, so the autocatalytic reactions continuously carry out the transformation of the initial reactants (R) to products (P) in the reactor (Figure 5.2), allowing the destabilization of the racemic state ($X_L = X_D$) in favor of one of the two nonracemic states. This process will occur as long as the chemical potential (μ) of the reagents (μ_R) is greater than the chemical potential of the products (μ_P), and this is accomplished by a continuous inward flow of achiral reagents and outward flow of products. This maintains chemical potentials $\mu_R > \mu_P$, therefore $\Delta G = \mu_P - \mu_R < 0$, and the autocatalytic reactions keep moving in the desired direction (R → L + D → P) (Figure 5.2a).

In a closed system, the decrease in concentration of R (consumption of reagents) and increase in concentration of P (accumulation of final products) will reach equilibrium: $\mu_P = \mu_R$, $\Delta G = 0$ (Figure 5.2b). At equilibrium, all components achieve identical chemical potential, $\mu_L = \mu_D$, including the intermediates L and D, and the system is therefore racemic ($X_L = X_D$). Closed systems are most commonly encountered in the laboratory, but there is an exception in the case of irreversible autocatalytic reactions. In this case the reaction does not actually reach a steady state, but due to its irreversibility accumulates one enantiomer at a time, and stops when the reagents are exhausted. This results in a highly stochastic behavior, with imperfect homochirality and large run-to-run fluctuations. Soai's reaction (Section 5.2.4, *Asymmetric Autocatalytic Reactions*) and certain crystallization experiments (Section 6.2, *Spontaneous Symmetry Breaking in Crystallization*) are examples. In these, kinetic factors prevent equilibrium being reached, and enantiomeric excess persists at the end of the reaction in a permanent metastable condition.

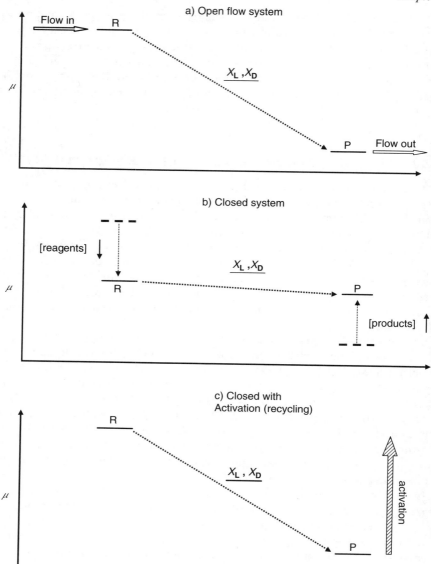

Figure 5.2 (a) An open-flow reactor, such as the original Frank model, with inward flow of reagents and flow outward of achiral products. This maintains ΔG stationary inside the reactor and autocatalytic reaction operating; (b) A closed system of reversible reactions lowers the free energy of the system by equilibrating the chemical potential of reagents and products, until a racemic equilibrium is reached ($\Delta G = 0$); (c) Alternatively, reactions in the closed system can be maintained working in the direction R → L + D → P by recycling of the products as reagents, for which an external energy source is necessary. (Adapted from Reference 6, with authorization.)

Strictly speaking, from a thermodynamic point of view an open-flow reactor is not mandatory for maintaining a system of autocatalytic reactions moving in the right direction, but it is a plausible solution. Another possibility is the recycling of products back to reagents (Figure 5.2c), within a closed but energetically non-isolated system. In this case the final products, P, rather than being flushed out of the reactor, are activated by an external source of energy and regenerated as reagents.[10]

This last modified Frank model shows a similar behavior, namely the appearance of steady nonracemic states in a non-equilibrium regime, driven by an incoming energy source. One such model involving prebiotically relevant molecules has been described.[11] This is based on cycles of polymerization–depolymerization of amino acids (or networks), in which the epimerization of the α-carbon of an incoming amino acid of opposite stereochemistry to those in the main peptide chain is stereoselective, favoring the formation of homochiral peptides. This is the key reaction step, playing the role of a formal autocatalytic step. The polymerization step is an endergonic process ($\Delta G > 0$) in aqueous media and requires activation of the amino acids by means of an external source of energy. This can be achieved by transfer of chemical energy from a set of external reactions that generates, for instance, the activated N-carboxyanhydrides of amino acids,[12] allowing the polymerization–depolymerization cycles to continue with the emergence of stationary chiral points. Other models based on Frank's model of autocatalysis have also been reported,[13] involving irreversible[14] and reversible polymerizations[15] in an open-flow system, or diverse recycled systems.[16]

5.2.3 Autocatalytic Amplification of the Weak Force

A deterministic mechanism for the origin of biomolecular homochirality can be devised by including the weak force as a mirror-symmetry breaking element and an autocatalytic mechanism—an extremely efficient one—to amplify its vanishingly small effects. The probability of a symmetry-breaking bifurcation such as that described by the Frank model (Section 5.2.1, *The Frank Model*) can be further assessed by considering the time (t) as a variable, representing the time under which a small chiral influence is acting on the system. The presence of a small bias that favors one of the bifurcating branches and the probability of the racemic system making a transition to the favored branch is described by an equation in which $d\alpha/dt$ is expressed in terms of concentration of the reactants, reaction rates, random fluctuations and systematic chiral influences.[17] Detailed examination of this equation shows that the sensitivity of the bifurcating process to the systematic chiral influence, depends on the rate at which the system moves through the critical point λ_C. This passage could correspond, for example, to a slow increase in the concentration of biomolecules in the primitive oceans.

Within the context of Frank's autocatalyzed reactions, in 1985 D. K. Kondepudi arrived at a computer simulation of an open flow reactor system suitable for a prebiotic scenario. The variables chosen to mimic such a scenario were as follows:

- A reactor of 4×10^6 L, equivalent to a small lake 1 km wide $\times$ 4 m deep.
- An increase in the concentration of the achiral reagents A and B ($\Delta[A]$ and $\Delta[B]$) from 10^{-3} to 10^{-2} M, during which the critical point is passed.
- A lapse of 10^4–10^5 years in the original metastable conditions under the effect of a small chiral bias (such as ΔE_{pv}).
- Realistic reaction rates of 10^{-10} mol L^{-1} s^{-1}.

Using this model, it is possible to understand the extraordinary sensitivity of the system to small systematic biases favoring one enantiomer by increasing its production rate. Slow passage through the critical point makes the system sensitive to very weak but systematic chiral influences. It can be estimated that, with a ΔE_{pv} between enantiomers of 10^{-17} kT, an amplification time of 10^4–10^5 years is required to select the favored enantiomer with a 98% probability. Increasing the magnitude of ΔE_{pv} can dramatically reduce the timescale and the reactor volume. The model also adapts to other features, such as competing racemization of the chiral products, and random fluctuations, larger than ΔE_{pv} but noise-averaging to zero over long time-scales and large volumes of reaction. As long as the autocatalytic production rate of chiral molecules is faster than racemization rates, the chiral symmetry-breaking Scenario (b) (Figure 5.1) is viable.[18]

This model has been carefully formulated and is highly heuristic, although its significance is uncertain given its exclusively theoretical basis. At present there is no experimental evidence to confirm such a scenario, involving the chemical reactions of prebiotic molecules and yielding enantiomerically enriched crudes—or at least, not on the human timescale.

There are, however, some parallels. The prediction of amplification has been verified in electronic devices, in which certain noise-driven nonlinear systems show an analogous behavior.[18] Closer to the chemical world, this also seems to be the case in the formation of chiral helical aggregates by slow concentration during stirring of achiral building blocks.[19] In this example, vortex motion—a minor physical chiral influence—determines the sense of the helix (Section 4.4, *Fluid Dynamics: Vortex Motion*).

5.2.4 Asymmetric Autocatalytic Reactions

Asymmetric autocatalysis in organic reactions was discovered by K. Soai in the addition of dialkylzinc to a number of structurally related aldehydes.[20] It is defined as an enantioselective synthesis, in which the chiral product acts as an asymmetric catalyst for its own production (Scheme 5.2).

Asymmetric autocatalytic reactions amplify the chirality of the reaction products compared to the starting materials or catalyst employed, therefore displaying what are called nonlinear chiral effects. They represent, however, a different case to other reactions also displaying nonlinear effects in which the

Mechanisms of Amplification

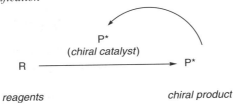

Scheme 5.2 In asymmetric autocatalytic reactions the chiral product and the chiral catalyst are the same compound, both in structure and chiral configuration.

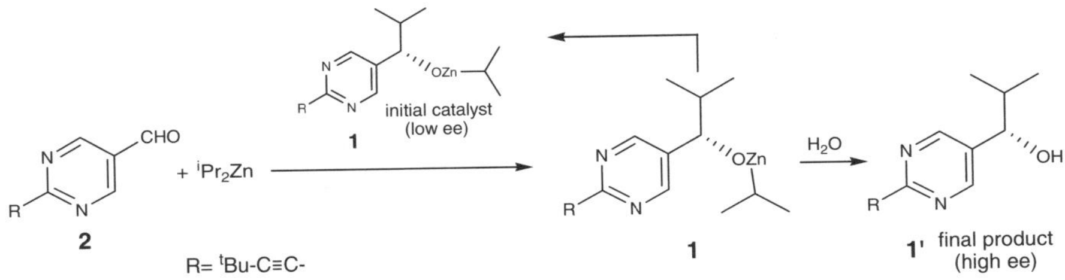

Scheme 5.3 High chiral amplification factors can be attained in the asymmetric autocatalyzed formation of 1-(2-alkynylpyrimid-5-yl) alkanolates (**1**) by addition of $^{i}Pr_2Zn$ to the 2-alkynylpyrimidine-5-carbaldehyde (**2**). Starting from 10^{-5} ee in the initial catalyst (of identical structure to the reaction product), chiral products (**1'**) > 99.5 ee are obtained after three consecutive cycles of autocatalysis. $^{i}Pr_2Zn$ = di-isopropylzinc.

dynamic aggregative behavior of the enantioselective catalysts is the only factor responsible for the amplification. This last type of reaction will be covered in Section 5.3, *Amplification of Chirality in Other Chemical Reactions*. Instead, in autocatalytic reactions, the key to this rare but efficient process lies in the mechanism of autocatalysis (Section 5.2, *Autocatalysis* and Section 5.2.1, *The Frank Model*). In this process the catalyst and the reaction products are the same species, and reaction products are generated with an enantiomeric excess higher than that of the chiral catalyst itself, provided that a mutual inhibition mechanism between enantiomers is operative. In an effective autocatalytic system the catalyst must be capable of self-replication, but this activity must also be accompanied by suppression of the activity of its enantiomer, in what was referred to in the Frank model as mutual antagonism (Scheme 5.1, Reaction 5).

In practice, the outstanding characteristic of the asymmetric autocatalytic reaction is that it is able to amplify a small initial chiral bias from levels never previously seen. An actual example of an autocatalytic process showing remarkable amplification of chirality is given in Scheme 5.3. 5-Pyrimidyl alkanols are known to be highly enantioselective autocatalysts. In the optimal variant of the reaction, diisopropylzinc is added to 2-alkynylpyrimidine-5-carbaldehyde (**2**) in the presence of a low-*ee* chiral catalyst consisting of

1-(2-alkynylpyrimid-5-yl)-1-alkanolates of zinc (**1**), giving as reaction product the same compound (**1**) with highly enriched *ee*.[21]

This reaction scheme fits the scenario of the Frank model well. From a more specific mechanistic view there are several proposed models, with either dimeric[22,23] or trimeric[23b,c] catalysts forming part of the transition state, and satisfying the mechanistic requirements of a closed autocatalytic system.[24] Kinetic studies point towards a second-order dependence of the catalyst in the rate law, which implies dimeric catalytic species in addition to the reagents in the transition state.[22,25] The expected different catalytic activity and properties of these homo- and heterochiral dimers may well account for the mutual inhibition mechanism. In particular, if homochiral dimers displayed higher reactivity than heterochiral dimers, the effect would be the expected chiral amplification. From a quantitative point of view, the asymmetric autocatalytic reaction of (**1**) starting from an extremely low *ee* of the initial catalyst is illustrative.[26] In this example, when (*S*)-1-(2-*tert*-butylethynylpyrimid-5-yl)-2-methyl-propanol (**1'**) with only *ca.* 0.00005% *ee* (*S*-isomer : *R*-isomer = 50.000025 : 49.999975) was used as the initial asymmetric autocatalyst, the first reaction run gave (**1'**) with an increased *ee* of 57%. When the product was used again for a second run, the *ee* increased to 99%. The third asymmetric autocatalytic run using (**1'**) of 99% *ee* gave almost enantiopure (**1'**), *ee* >99.5%, which is virtually at the detection limit (*S* : *R* = *ca.* 400 : 1 by HPLC–UV). Taken together, during the three consecutive asymmetric autocatalytic cycles the proportion of the initially dominant *S*-enantiomer overwhelmingly exceeded that of the *R*-enantiomer. Significantly, the experiment showed reproducible effects when the sign of the initial chiral bias was reversed, *i.e.*, by using *R*-(**1'**) of 0.00005% *ee*. It should be noted that below the threshold of 10^{-5}% *ee* this asymmetric autocatalysis began to give signs of erratic behavior.[27]

The expected stochastic dispersion of a racemic mixture is given by the standard deviation of the sample size, $\sigma = \sqrt{N}/2$ (see Section 2.1.1.1, *The Racemic State*). For the above experiment (typical procedure: 0.013 mmol catalyst of 0.00005% *ee*), the intrinsic stochastic dispersion in the composition of a racemic sample of this size corresponds to fluctuations of $ee = \sqrt{N}/N = 3.5 \times 10^{-10}$. Interestingly, the reported minimum threshold of 10^{-5}% *ee* is *ca.* 3 orders of magnitude above the expected statistical standard deviation of the sample. In other words, the reaction is amplifying a much larger effect, also of a stochastic nature, but different from the statistical deviation from 1 : 1 in the composition of a racemic sample. This unknown effect may be related to small fluctuations produced in the initial stages; in this case a study of the effect of reactor size and rate of homogenization may provide clues about the nature of the observed random behavior.

Without the addition of any chiral catalyst, a mirror-symmetry breaking scenario should be operative. Indeed, indications of spontaneous symmetry breaking for an analogous reaction to that described in Section 5.3 followed immediately. In the absence of any chiral substance or influence (*e.g.*, chiral catalyst), the reaction described in Section 5.3—when run in diethylether—gave enantiomerically enriched pyrimidyl alkanolate (**1**) with a random outcome in the configuration of the chiral center.[28] From 37 runs, statistical analysis showed a bimodal distribution of products (19 times *S* and 18 times *R*), indicative of spontaneous symmetry breaking, similar to that obtained in the crystallization of $NaClO_3$ (Section 6.2, *Spontaneous Symmetry Breaking in Crystallization*). Confirmation of the strictly random outcome of this reaction provides the first example of a chemical reaction exhibiting spontaneous symmetry breaking. The reaction outcome is apparently the effect of random molecular fluctuations, perhaps from the initial stages of the reaction, increased by a self-amplifying autocatalytic process (Section 5.2, *Autocatalysis*). Interestingly, a divergence from random distribution has been observed by the authors for this same reaction when run in toluene, the consequence of unknown chiral effects. Some systematic statistical deviation was also found in the absence of chiral additives in a similar reaction run by independent researchers.[29] This systematic effect was attributed to chiral impurities in the solvent (ultimately of biogenic origin), impurities which defied every attempt at detection; this demonstrates the extreme sensitivity of this reaction to chiral influences. Another example of spontaneous symmetry breaking has been reported in the presence of achiral silica gel.[30]

A wide variety of chiral initiators other than (**1**) can be used in this reaction, including not only a range of polar chiral molecules, but also chiral hydrocarbons such as hexahelicene,[31] allenes[32] and binaphthyl,[33] or cryptochiral saturated quaternary hydrocarbons having no functionality.[34] Also, the presence of chiral inductors at a non-molecular level, such as CPL (circularly polarized light), inorganic chiral material such as quartz or $NaClO_3$ crystals, chiral organic crystals comprised of achiral molecules, among others, are chiral influences which can successfully trigger the formation of (**1′**) in high *ee* (Scheme 5.4 and Table 5.1).[35]

From Table 5.1 it is seen that asymmetric autocatalysis provides a unique method for the discrimination of many kinds of chiral influences. In the case of circularly polarized light (CPL), the process represents an improvement in the correlation between the chirality of the CPL and the *ee* of an enantioselective chemical synthesis initiated by CPL (for a comparison with CPL-driven chemical reactions, see Section 4.3, *Asymmetric Photolysis and Photosynthesis*).

At this point it is of interest to mention physical forces which have failed to induce a chiral sign in an achiral chemical system. This has been the case with

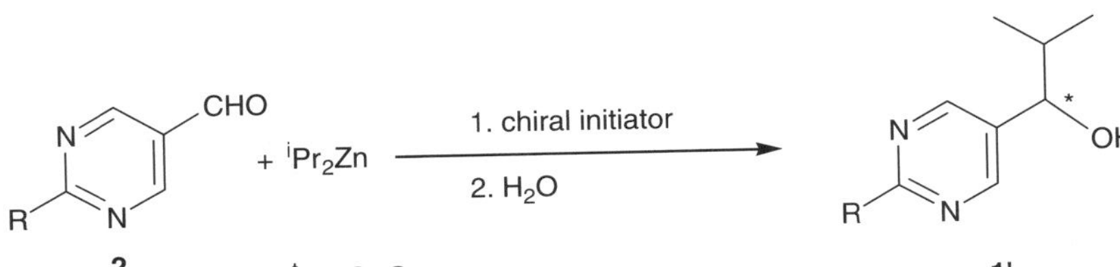

Scheme 5.4 A wide range of chiral initiators different from the typical molecular chiral catalysts can induce a large *ee* in Soai's reaction. Some of these chiral initiators are collected in Table 5.1.

Table 5.1 The effect of different chiral initiators in the reaction of Pr$_2^i$Zn with aldehydes (2) (Scheme 5.4).[36]

Chiral influence	Handedness	Product (**1'**) (% ee, and configuration)	Reference
Circularly polarized light (CPL)	r	>99.5 (R)	37
	l	>99.5 (S)	37
Quartz	d	97 (S)	38
	l	97 (R)	38
NaClO$_3$	d	98 (S)	39
	l	98 (R)	39
NaBrO$_3$	d	98 (R)	39,40
	l	97 (S)	39,40
Artificial helical silica	Left-handed	97 (R)	41
	Right-handed	96 (S)	41
Silsesquioxane	(R,R)	93 (R)	42
	(S,S)	95 (R)	42
Ephedrine on silica	(1R,2S)	97 (R)	43
	(1S,2R)	95 (S)	43
Chiral organic co-crystals (tryptamine/4-Cl-benzoic acid)	P	96 (R)	44
	M	95 (S)	44
Chiral organic co-crystals (phenanthridine/3-indole-propionic acid)	P	92 (S)	44
	M	97 (R)	44

r, l = right handed, left handed; d, l = dextrorotatory, levorotatory; P, M = stereodescriptors *plus*, *minus*.

β-radiation, a truly chiral influence originating from β-radioactive emitters, which has defied all attempts to imprint its sign within chemical processes (Section 4.2.6, *β-Radiolysis,* and Section 4.2.2, *The β-Decay*).

5.2.5 Self-replication

Self-replication is the process by which some molecules are able to generate exact copies of themselves. Not surprisingly, this is of special interest in the case of biomolecules: within the context of the definition of life self-replication plays a central role in theories relating to the origins of life. In a self-replicating process, a molecule acts as a template for reproducing components or modules of its structure, bringing them together and arranging them in space by non-covalent interactions. Once collected in the proper orientation, these can readily bind to one another. A new copy of the original molecule emerges from the original template in a type of modular self-assembly. Formally, this is another case of autocatalysis (Section 5.2.1, *The Frank Model*), the name being reserved for special molecules, mostly nucleotides and peptides, in which molecular recognition plays an important role.

Some of the first studies of self-replication used complementary oligo-nucleotides and related molecules.[45] These studies concentrated on the

Mechanisms of Amplification

template-based mechanisms of self-replication of oligonucleotide strands of natural chirality.[46] The replication of a polynucleotide is an obvious example of such autocatalytic growth. A selected homochiral sequence of polynucleotides could replicate by directing the polymerization of monomers of similar handedness, while excluding monomers of the opposite handedness. However, in practice this is not without its problems. In template-directed reactions involving a racemic mixture, monomers of the opposite handedness to the template become incorporated as chain terminators.[47] Other studies indicate that enantiomeric cross-inhibition is problematic in the polymerization of nucleotides on RNA and DNA, as well as with other templates.[48] More recently, self-replicating peptides have also arrived on the scene.[49]

Using a peptide based on the leucine-zipper motif of the yeast transcription factor GCN4, M. R. Ghadiri, et al., demonstrated the feasibility of peptide replication. Figure 5.3 illustrates the self-replication process, beginning with the 32-residue peptidic template (in grey) which holds together a 15-residue nucleophilic fragment (in red) and a 17-residue electrophilic fragment (in blue) and catalyses the thioester-promoted amide condensation bond (in yellow), hence catalyzing its own formation. The self-replicating process displays parabolic growth and is governed by square-root kinetics (i.e., the rate of the autocatalytic reaction is proportional to the square root of the concentration of the template) during the initial period of product formation.

Later, Ghadiri's group demonstrated that these self-replicant peptides were capable of efficiently amplifying homochiral products from a racemic mixture of peptide fragments through a chiroselective autocatalytic cycle. In addition, the amplification process discriminated between structures possessing only a single stereochemical mutation, i.e., peptides with only one D-aminoacid

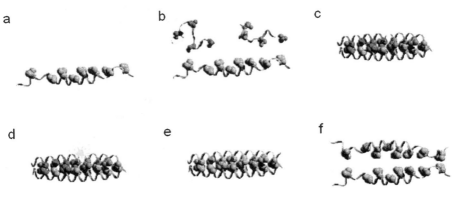

Figure 5.3 Self-replication. A self-replicating peptide (a) or "replicator" is linked non-covalently with two smaller peptide fragments (b, c). Reorientation of the fragments allows them to combine covalently (d) to generate a copy of the initial peptide (e), which splits in the medium into two replicas (f). The replicator is capable of self-generation when fed with a racemic mixture of peptide fragments, thereby amplifying the chirality of the product. (Pictures courtesy of M. R. Ghadiri, personal communication).

within an otherwise homochiral sequence, in the nature of a dynamic error-correction function.[50]

Other systems of peptides showing self- or cross-catalytic activity have been designed.[51] Attributes of living systems such as chiroselectivity, dynamic error correction, response to changes in environmental conditions and cross catalysis are exhibited by various self-replicant peptide systems. These results support the concept that self-replicating polypeptides could have played a role in the origin of homochirality on Earth, in the same way as polynucleotides have been traditionally considered the archetype of self-replicant molecules. Although free amino acids do not condense spontaneously to produce peptides in aqueous media, a number of activations are known which could have operated in a prebiotic environment.[52] The salt-induced peptide formation reaction (SIPF reaction) is particularly illustrative.[53] Sodium chloride present in increasing concentrations as the result of evaporation acts as a dehydrating agent, while the presence of Cu(II) acts as a catalyst and reaction center by complexation, driving the formation of peptides from amino acids in evaporating aqueous solutions.

5.3 Amplification of Chirality in Other Chemical Reactions

Asymmetric amplification occurs in a chemical reaction when reaction products are obtained in higher enantiomeric excess (*ee*) than the chiral influence, for example the chiral auxiliary or catalyst, as represented in Scheme 5.5. This may or may be not the consequence of an autocatalytic process, but in either case a nonlinear effect will be observed (Section 5.3.1, *Nonlinear Effects*). The autocatalytic case has been considered separately (Section 5.2.4, *Asymmetric Autocatalytic Reactions*).

5.3.1 Nonlinear Effects

H. B. Kagan and collaborators were the first to study this effect in a range of organic synthetic processes, including the asymmetric oxidation of methyl *p*-tolyl sulfide, asymmetric epoxidation of geraniol in the presence of various chiral titanium complexes, as well as the proline-catalyzed Hajos–Parrish reaction.[54] Prior to these studies it had been generally assumed that a chiral auxiliary would produce chiral induction levels in reaction products either lower than, or at most

Reagents $\xrightarrow{\text{Cat* } (\%ee_{aux})}$ Products*

asymmetric amplification: $ee_{(prod)} > ee_{(aux)}$

Scheme 5.5 Asymmetric amplification occurs when reaction products are obtained in higher *ee* than the chiral influence.

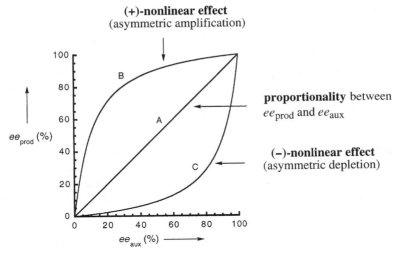

Figure 5.4 Nonlinear effects can be demonstrated in asymmetric synthesis by plotting the ee_{aux} of the chiral auxiliary vs the ee_{prod} of the products obtained. In a chemical reaction displaying nonlinear effects there may be an amplification of the expected ee, depletion of ee, or a complex relationship between ee_{prod} and ee_{aux}. These asymmetric amplifications require an initial scalemic (nonracemic) mixture to produce the amplification. (From Reference 57, with authorization.)

equal to, those of the chiral auxiliary. This is an intuitive rule which holds in most instances, provided there is no aggregative behavior in the catalyst.

The relationship between the ee value of the chiral auxiliary and that of the product deviates from linearity in a number of reactions, to give what are described as nonlinear effects (Figure 5.4).[55] Nonlinear effects in asymmetric synthesis and stereoselective reactions are the result of molecular interaction and complexity beyond the ideal behavior of the solutes. This is displayed in the degree of association of the chiral catalyst and the effect of the aggregates on the reaction mechanism and the products formed. To gain a better understanding of the behavior of chiral catalysts during reactions leading to nonlinear effects a number of mathematical models have been developed,[56] and within this frame, description of the actual reactions involved have been reviewed.[57] Although these models were initially devised for organometallic catalysts, with which aggregation processes are more often present, they are equally applicable to any chiral catalyst.

Starting from a scalemic (enantioenriched) chiral catalyst, the dynamic behavior of the chiral catalyst can generate new species (aggregates), such as homo- and hetero-dimers, trimers, and oligomers of various compositions, in the reaction media involved in complex equilibria. Disregarding unreactive species, each of the remaining active catalysts may display better, lower, opposed, or no enantioselectivity at all in their reactions as compared to the enantiopure catalyst. The final ee observed will be the contribution of all the competing catalyzed reaction pathways, and the final outcome might be amplification, depletion or changing behavior, depending on the *ee* of the catalyst. This amplification scenario shares certain aspects with those describing the amplification of scalemic mixtures through classical physical processes, such as dissolution or sublimation (Section 5.6, *Amplification of Scalemic Compounds: Eutectic Mixtures*). It is the precise nature of the intermolecular interactions in the scalemic catalytic species that will determine the prevalence of one or another type of aggregate, which later may be either conducive or not to chiral amplification. And it is subject to the same restrictions, such as necessarily originating from scalemic mixtures—in other words, these reactions do not promote mirror-symmetry breaking, but amplify a pre-existing enantiomeric chiral bias. This very crucial point is what makes true *Asymmetric Autocatalytic Reactions*, considered earlier in Section 5.2.4, distinctly different.

5.3.2 Amplification of Chirality by Cooperative Forces: Growing Polymers and Supramolecular Assemblies

The assemblage of small molecules into macromolecular or supramolecular structures, either through covalent bonding or by means of non-covalent interactions, both in a cooperative manner, may lead to chiral amplification. Examples of each type are known. These and similar cooperative interactions displaying nonlinear chiral effects certainly may have played an important part in the development of single-handedness during biochemical evolution.

5.3.2.1 Amplification of Chirality in Polymerizations

Amplification of chirality during polymerization is a relevant factor which could have been involved in the evolutionary stages prior to the appearance of fully homochiral biological structures. The case of amino acids activated as *N*-carboxyanhydrides of amino acids (NCA-amino acids) has long been recognized.[58] The polymerization of NCA-amino acids is triggered by a nucleophilic initiator, producing peptide chains and releasing CO_2. Polymerization of NCA-amino acid racemates possessing hydrophobic chains (tryptophane, leucine and isoleucine) in aqueous solution yielded oligopeptides characterized by a high degree of homochirality in their sequences.[59] When the polymerization was carried out in water using an enantioenriched mixture of monomers (20% *ee* of NCA-L-leucine, or 20% *ee* of NCA-L-glutamate), chiral amplification of the homochiral oligopeptide fractions was detected up to 73% for $(Leu)_5$, and of 71% for homochiral $(Glu)_7$.[60,61] Studies of the cooperativeness of the polymeration showed that the probability of the addition of either a L- or a D-amino acid monomer to the growing end depended on the configuration of the last two monomeric units of the growing oligomer end (or second-order Markov process),[62] as well as by cooperative amplification resulting from formation of an α-helix.[59]

There is no doubt that the secondary structure of polymers is involved in the amplification of chirality in many polymerizations. A notable case is in the polymerization of isocyanates (R–N=C=O) to produce poly(isocyanates), [–(R)N–C=O–]$_n$, a polymer known as Nylon-1.[63] Nylon-1 displays an unusually strong preference for local helical conformation of its chain. This polymer, although helical, can be synthesized from achiral monomers (*e.g.*, R = *n*-hexyl), yielding equal numbers of right- and left-handed helical chains. However, the polymer is unusually sensitive to chiral influences. The introduction of a minor chiral bias by introducing a small amount of a chiral isocyanate (*e.g.*, R = (*R*)-2,6-dimethylheptyl) can induce a high enantiomeric excess (*ee*) of one helical sense. Thus, a copolymeric chain with only 15% of this chiral monomeric unit and 85% of *n*-hexylisocyanate gives the same degree of optical activity as a chain composed entirely of chiral units.[64] This type of chiral amplification, consisting of generating an excess of the preferred helical sense by the influence of a small number of chiral units (the "sergeants") over a large number of achiral units (the "soldiers"), is known as the sergeants-and-soldiers effect.

Another example is a polyisocyanate synthesized by the random copolymerization of a racemic mixture of monomers (containing only a slight excess of one enantiomer), which shows a large excess of the helical form generated from homopolymerization of the corresponding enantiopure monomer. For example, a polymeric chain made from a monomer of only a 12% *ee* of (*R*)-2,6-dimethylheptylisocyanate] gives a CD (circular dichroism) spectrum identical to that of a polymer synthesized from almost enantiopure monomer. This generation of an excess of the helical sense preferred by the excess enantiomer is called the "majority rules" effect, from the minority component taking part of the helical sense imposed by the majority units.[65]

Another experiment showing the strong cooperative response to chiral information of this helical polymer is the polymerization of (*R*)-1-deuterio-*n*-hexylisocyanate, [α]$_D$ = +0.65 (neat), to give the corresponding poly(isocyanate) with optical activity [α]$_D$ = −444 (hexane). The sole isotopical substitution of a hydrogen atom by deuterium in the *n*-hexylisocyanate monomer in an enantioselective manner, triggers the almost complete prevalence of a preferred helical sense in the polymer.[66] The amplification mechanism giving rise to the helical excess has been interpreted as a conformational equilibrium isotope effect, displaying an extremely efficient additive and cooperative mechanism acting in the polymeric chain. Whether accurate or not, this is an example of the extreme sensitivity of this system towards chiral bias. The amplification of chirality in other helical polymers, such as polyphenylacetylenes, polyisocyanides and polysilanes has also been studied, both experimentally and to some extent theoretically.[67]

5.3.2.2 Amplification of Chirality in Supramolecular Assemblies

The transfer of chiral information to achiral or dynamically racemic supramolecular systems from chiral guest molecules through non-covalent interactions can be used for determining the sense and amplifying the chirality in a range of systems. This has been applied to many chiral transitions in the liquid crystal phase, where successful assembly of large supramolecular structures depends on a delicate balance between non-covalent forces and dynamic structural interactions. The sensitivity to small chiral perturbations in some of these systems has long been recognized, and is often displayed in an amplified way throughout the whole supramolecular arrangement.[68] The supramolecular structure usually takes on a helical geometry, amplifying the initial chiral bias.[69] The mechanism of amplification shares certain characteristics with formal phase transition and can be described as follows. In nematic phases, the individual molecules are randomly distributed, but intermolecular interactions tend to orientate them in a parallel arrangement to give an arbitrary common axis, or director. A small amount of a chiral initiator (or chiral dopant) can induce in a nematic phase, composed of achiral molecules and therefore having centrosymmetric (achiral supramolecular) structure, a new kind of helical organization—a cholesteric phase (chiral supramolecular organization). In cholesteric phases, the oriented molecules are organized in horizontal layers, each slightly rotated in relation to its neighbor, forming columns with a helical suprastructure. The resulting assembled chiral motive can be characterized in terms of its pitch, which is usually tens or hundreds of times larger than the constituent molecules, and its helicity, which depends on the chiral sense of the inductor molecule. The amount of chiral dopant needed to induce a cholesteric phase may be very low (*e.g.*, 10^{-4} molar fraction or lower).[70]

These cooperative interactions are analogous to the sergeant-and-soldiers effect (Section 5.3.2.1), but acting on non-covalently linked assembles and displaying in many cases an even larger amplification of the chirality. In a similar fashion, chiral amplification in the form of the majority rules effect is also widely found within this family of compounds. Many of them are disk-shaped molecules, capable of adopting low-energy chiral conformations by complexing the dopant, and hence configuring the platform on to which the remaining molecules can dock (see also Section 4.2, *Fluid Dynamics: Vortex Motion*).[71] The exact nature of the underlying forces operating in these assembles is complex and empirical, and for the moment its interpretation is beyond the reach of a theoretical study. A better understanding of the amplification of chirality in such dynamic aggregates is needed; this may provide simple mechanistic routes for generating homochiral biopolymers from almost racemic monomers, provided the monomer undergoing the polymerization is in a well-defined self-assembled state.

5.4 The Yamagata Cumulative Mechanism

Another amplification mechanism is the Yamagata cumulative mechanism,[72] historically one of the first proposed links between ΔE_{pv} (Section 4.2.5, *Quantification of the Parity Violation in Molecules: ΔE_{pv}*) and the appearance of asymmetric biomolecules on Earth. This mechanism implies the additive

Mechanisms of Amplification

accumulation of the ΔE_{pv} of individual building blocks when these assemble to form a macromolecular structure, for example a polymer or a crystal. In its thermodynamic version, it is implicit to this mechanism that the parity-violating energy difference of the macromolecule (consisting of n building blocks) would be n times that of an individual building block, resulting in a greatly amplified effect at the macroscopic level. In the instance of an α-quartz crystal of given handedness, consisting of n distorted [SiO$_2$] tetrahedrons, ΔE_{pv}, of the crystal would be n times ΔE_{pv} of a single distorted [SiO$_2$] tetrahedron, *i.e.*, ΔE_{pv} ([SiO$_2$]$_n$crystal) = $n \cdot \Delta E_{pv}$([SiO$_2$]).

The theory was first developed for reactions under kinetic control, *e.g.*, in growing polymers or crystals, for which similar considerations applied to the activation energy.[73] Indeed, for an unlimited supply of racemic monomer, the rate at which a polymer composed of n L-monomers is formed, relative to its enantiomeric D-polymer (N_L/N_D), would be: $N_L/N_D = e^{n\Delta E_{pv}/kT} \cong (1 + \Delta E_{pv}/kT)^n$, in which last term is simplified using the Maclaurin series expansion of the exponential function. The expression displays a cumulative effect of $n\Delta E_{pv}$ in the Arrhenius activation energy difference between the two polymerizations (*i.e.*, the degree of asymmetry in the polymers increasing with the degree of polymerization), ΔE_{pv} standing for the parity-violating energy difference between the two enantiomeric monomers. In the case of polymerizations, recent *ab initio* calculations of E_{pv} involving polypeptides of up to 8–11 monomers (Ala$_8$ and Gly$_{11}$) indicates that, although the total value of ΔE_{pv} is the sum of the monomeric contributions, there is no additional effect caused by the secondary α-helical structure of these molecules.[74] If extrapolated to the polymerization reaction of larger proteins (such as those conceivable in a prebiotic environment) the net effect of E_{pv} may be considered negligible compared to the standard deviation due to statistical fluctuations (Section 2.2.1.1, *The Racemic State*). To yield a measurable E_{pv} difference under these conditions, an extremely large and unrealistic number of monomer units (of the order of 10^{13-14} units) would be required, as well as a perfect additivity of the single contributions.

The main drawback of the Yamagata cumulative mechanism arises in the study of crystallization, in particular of α-quartz. The cumulative mechanism simply does not hold in this situation. α-Quartz is a covalent enantiomorphic crystal which can be considered as a polymer of distorted tetrahedral [SiO$_2$] units (structural details and conventions on terminology are explained in Section 8.2, *Chiral Crystals and Faces on Crystals*). From published calculations (Section 4.2.5, *Quantification of the Parity Violation in Molecules:* ΔE_{pv}), E_{pv} for *l*-quartz was found to be stabilized by *ca.* 10^{-17} kT per [SiO$_2$] unit with respect to *d*-quartz.[75] Under thermodynamic conditions, *i.e.*, the equilibrium expressed by Equation (5.8), an amplification of $n\Delta E_{pv}$ in the energy of a *l*-quartz crystal compared to a *d*-quartz crystal, both constructed of n [SiO$_2$] units, would produce an expected *ee* equal to that shown in Equation (5.9), where $\Delta E_{pv} = 2E_{pv}$; this was derived from the definition of enantiomeric excess (*ee*):

$$[d] = [l]e^{-n\Delta E_{pv}/kT} \tag{5.8}$$

$$ee(l) = \frac{[l] - [d]}{[l] + [d]} = \frac{1 - e^{-n\Delta E_{pv}/kT}}{1 + e^{-n\Delta E_{pv}/kT}} = \tanh\left(\frac{nE_{pv}}{kT}\right) \quad (5.9)$$

Numerically, for any crystal of α-quartz of prismatic habit of *ca.* 1 mm (*ca.* 0.1 mg, $d_{\alpha\text{-quartz}} = 2.648 \text{ g cm}^{-3}$), the calculated *ee* favoring left-handed crystals is virtually unity: $ee_{(\text{L-crystals})} = 0.999\,999\,996$. This would overwhelmingly favor the occurrence of untwinned *l*-quartz specimens of such size or larger, provided the Yamagata accumulative mechanism was applicable. This does not however seem to be the case (Figure 5.5). On the other hand, if the occurrence of quartz crystals in nature was controlled by nucleation or any other kinetically controlled processes,[76] Equation (5.8) would not hold. In such a case, a statistical sampling of quartz from different locations throughout the world would display an observable preference for the *l*-quartz. This also is not the case.

Figure 5.5 A large crystal of *d*-α-quartz, right-handed, just like the hand of one of the authors holding it (A.G.). According to Equation (5.9), derived from the application of the Yamagata cumulative principle (under conditions of thermodynamic equilibrium), crystals of *d*-quartz of such a size—or even much smaller than this—would be such a precious rarity that they simply would not be in our hands.

Analysis of the distribution of enantiomorphic right- and left-quartz throughout the world shows that *l*- and *d*-quartz crystals are distributed in equal amounts in every location.[77] The presence of optically active *l*-quartz crystals or other silicate minerals was at one time proposed as the source of homochirality in the evolution of the biosphere, mainly based on a reported small excess of left-quartz crystals (1% *ee*) on Earth.[78] This supposed excess was later proved to be statistically inconsistent, and consequently the Yamagata amplification in the calculated ΔE_{pv} seems inapplicable in its current version. The features of this theory rest on weak grounds and have been radically refuted[79] due to lack of experimental validation, although sporadic support is also found in crystallization.[80] See also Section 4.2.3.1, *Experimental Confirmation of Parity Violation in Weak Interactions*.

5.5 The Salam Phase Transition

Postulated by Nobel Prize winner A. Salam, a third and unique amplification mechanism states that ΔE_{PV} between enantiomers might lead directly to homochiral products through an abrupt second-order phase transition involving a quantum mechanical Bose–Einstein condensation.[81] If so, the electroweak interaction might promote tunneling through the potential barrier existing between the two enantiomers of a chiral molecule. This tunneling, or Salam phase transition, would be effective below a critical temperature, T_c, in which case an equilibrium shift towards the enantiomer having the more stable ΔE_{PV} might occur. T_c is necessarily very low, perhaps only a few tenths of K, which lies in the range of interstellar temperatures. Attempts to validate this theory by freezing down to 0.1 K a sample of racemic cystine failed, giving no change in optical rotation—as many would in fact have predicted.[82] Recent claims reporting direct observation of the Salam phase transition in amino acids at much higher temperatures by differential scanning calorimetry (DSC), magnetic susceptibility, Raman spectroscopy,[83] and NMR experiments[84] were quickly contested by a separate group after reexamination of these experiments, as well as X-ray diffraction studies.[85] Later, the possibility of L–D stereomutation was definitively refuted.[86] This mechanism does have a physically sound basis but it seems to be kinetically hindered, precluding a Bose–Einstein condensation from racemic to the favored ΔE_{pv} enantiomer. Just like the previous mechanism, it remains speculative and has so far not been confirmed experimentally.

5.6 Amplification of Scalemic Compounds: Eutectic Mixtures

A mechanism of amplification between different phases can be anticipated from the properties of scalemic (scalemic = enantioenriched) mixtures when subjected to classical physical processes such as dissolution–crystallization or sublimation–deposition. Indeed, rather than an amplification mechanism, it could be considered a partitioning or a fractionation process,[87] in which different phases are involved and an unequal distribution of enantiomers occurs between them.

The underlying basis on which a scalemic mixture can fractionate its enantiomeric components lies in the nature of the intermolecular forces between molecules. In simple terms, there may be homochiral interactions (interactions between molecules of the same chirality, say, L–L=D–D) and heterochiral interactions (interactions between enantiomeric molecules, L–D=D–L). These interactions are diastereomeric, and this nonequivalence in energy is particularly apparent in the solid state, although not exclusive to it. Regarding the relative strength of these interactions, a racemic mixture can crystallize in one of three ways:[88]

- *Conglomerates*, which are a mechanical mixture of enantiomerically pure crystals of each enantiomer. In this case homochiral molecular forces prevail in the solid state. The melting point of the racemic conglomerate is lower than the pure enantiomer, as deduced from evaluation of the chemical potentials for each enantiomer in each phase, and is the smallest possible among all the scalemic mixtures. A conglomerate of enantiomorphous and hemihedral crystals was observed by Pasteur in the crystallization of racemic sodium ammonium tartrate as the tetrahydrate, when conducted from aqueous solution below 27 °C. Organic compounds crystallizing as conglomerates account for *ca.* 10% of all crystalline organic compounds, and occur in any of the 65 chiral space groups.
- *Racemic compounds* (or *true racemates*), in which the molecules form a single crystalline phase with the two enantiomers present in an ordered 1:1 ratio in the elementary cell (*i.e.*, in a racemic crystal). In this case, heterochiral molecular forces predominate in the solid state. By the addition of small amount of one enantiomer to the racemic compound, the melting point decreases, again as deduced from evaluation of the chemical potential in each phase, although the pure enantiomer can have either higher (more seldom) or lower (more common) melting point than the racemic compound. Above 27 °C, Pasteur's racemic sodium ammonium tartrate crystallizes as the monohydrate, a holohedral racemate with a lattice in which the two enantiomeric tartrate ions share an inversion centre. This is the most common way in which enantiomeric mixtures crystallize, accounting for *ca.* 90% of all racemic mixtures of organic compounds, and these overwhelmingly crystallize in one of the 165 racemic (centrosymmetric) space groups.[89]
- Finally, *pseudoracemates* (sometimes called *racemic solid solutions*), which are solid solutions of both enantiomers in equal proportions in the crystal but which coexist in an unordered manner in the crystal lattice.[89] In pseudoracemates, homo- and heterochiral forces are very similar, and on addition of small amounts of one enantiomer to the racemic compound the melting point barely changes, or not at all. This type of crystal is relatively rare.

5.6.1 Solubility Properties

When a mixture of racemic molecules is crystallizing, a complex set of intermolecular forces operate, and a final balance is reached which in practice may be governed by both thermodynamic and kinetic parameters. It is sometimes observed that molecules packed in racemic space groups giving racemic compounds may be more favorable than if they are packed in chiral space groups, giving conglomerates. Among the numerous forces operating in crystal packing, the antiparallel alignment of molecular dipoles through an inversion centre allows mutual dipole cancelation. This factor accounts in part, although may not prevail, for the dominance of racemic compounds among organic crystallites. Discrimination between intermolecular stereoisomeric forces are manifested in the fusion, and also in the solubility and sublimation properties, of racemic and enantiopure compounds.

It has been known for a long time that for most chiral compounds there are significant differences in the solubility of the racemate and the corresponding enantiopure compound. For instance, in terms of heat of solution, Table 5.2 gives the enthalpies in solution of enantiopure solids, $D_{(s)} \leftrightarrows D_{(aq)}$ (1), or $L_{(s)} \leftrightarrows L_{(aq)}$ (2), racemic compounds, $DL_{(s)} \leftrightarrows DL_{(aq)}$ (3), and differences between the enthalpies of solution: ΔH_s (enantiopure) $- \Delta H_s$ (racemic), in water at 25 °C for tartaric acid and some amino acids.[90] Since the heat of mixing of the enantiomers in solution is almost zero, $DL_{(aq)} \leftrightarrows D_{(aq)} + L_{(aq)}$ (4),[91] ΔH_s (enantiopure) $- \Delta H_s$ (racemic) $= \Delta H_{mix} [DL_{(s)}]$ is an actual measure of the enthalpy of mixing of the opposite enantiomers in the solid state, according to the equation: $L_{(s)} + D_{(s)} \leftrightarrows DL_{(s)}$ (5). This is better illustrated in the following Hess cycle: $\Delta H_{mix}[DL_{(s)}] = \Delta H(5) = 1/2[\Delta H(1) + \Delta H(2)] - \Delta H(3) - \Delta H(4) = \Delta H(1) - \Delta H(3)$, provided that $\Delta H(1) = \Delta H(2)$, and $\Delta H(4) \cong 0$. For all the compounds in Table 5.2, apart from threonine, which has $\Delta H_{mix}[DL_{(s)}] \cong 0$, the formation of the racemic compound from its enantiomers in the solid state is exothermic. The occurrence of compounds with endothermic enthalpies of mixing is unlikely, although not impossible, corresponding to kinetically favored racemic phases. In addition to thermochemistry, for an adequate understanding of the melting or solubility properties of scalemic mixtures the phase diagram of the racemate–enantiomer system has to be constructed. Binary phase diagrams

Table 5.2 Enthalpies of solutions in water at 25 °C (kcal mol^{-1}).[a]

	ΔH_s (enantiopure)	ΔH_s (racemic)	ΔH_s (enantiopure) $- \Delta H_s$ (racemic)
Alanine	1.75 (L)	2.23 (DL)	-0.48 ± 0.05
Glutamic acid	5.86 (D)	6.84 (DL)	-0.98 ± 0.05
Histidine	3.25 (L)	3.61 (DL)	-0.36 ± 0.05
Threonine	2.34 (L)	2.30 (DL)	0.04 ± 0.03
Valine	0.686 (L)	1.23 (DL)	-0.547 ± 0.002
Tartaric acid	6.12 (+)	3.87 (±)	-2.25 ± 0.07

[a]Adapted from Reference 90.

describing the melting behavior of conglomerates, racemic compounds and pseudoracemates are well known.[88] These rectangular plots have axes showing temperature *vs.* the molar fraction of one enantiomer. But solubility relationships are more likely to be related to the prebiotic scenarios. For solubility properties, the inclusion of the solvent creates a ternary system. We shall briefly consider the main types of ternary systems describing the solubility of scalemic mixtures.

The solubility of a conglomerate:

For a three-component or ternary system, say, a conglomerate of $L_{(s)}$, $D_{(s)}$ and S (solvent), Gibbs's phase rule, $F + P = C + 2$ (F = degrees of freedom, P = phases, C = components) has $F = 3 - P + 2 = 5 - P$. To make a two-dimensional plot, we must keep two variables fixed, namely T and P, *i.e.*, isothermal and isobaric conditions. If we consider for instance a single-phase system under these conditions (*e.g.*, a specified unsaturated solution of L and D at constant pressure and temperature), $F = 3 - P = 2$, and the two variables will be the molar fractions of the two components. Once x_L and x_D are fixed, x_S is fixed. A rectangular plot is possible, with axes x_L and x_D, like a classical melting diagram. However, Gibbs suggested the use of an equilateral triangle plot, and this has become standard for ternary systems. From a given overall composition, *i.e.*, a point inside the triangle, the length of the perpendicular lines to the sides of the equilateral triangle gives the molar fraction of the three components. These lengths can easily be read by using the triangular coordinate system customarily used in ternary phase diagrams; all points in a line parallel to one side have an identical molar fraction of the corresponding component, sited in the opposite vertex. Consider now a saturated solution, *i.e.*, $L_{(s)}$ and $D_{(s)}$ in equilibrium with the solution, at fixed pressure and temperature. The isothermal–isobaric plot has $F = 3 - P = 3 - 3 = 0$; zero degrees of freedom indicates that the solution has a fixed composition for all components, which is their solubility at the fixed pressure and temperature. This is the eutectic point, and its composition is the eutectic composition. Since the solubility of $L_{(s)}$ must equal that of $D_{(s)}$, a conglomerate exhibits a solution with $x_L = x_D$, therefore $ee = 0\%$, regardless of the initial relative amounts of $L_{(s)}$ and $D_{(s)}$, as long as the three phases with both solids coexist. This means that no amplification occurs with conglomerates in solution. However, at this point, a trivial mechanism of amplification in the solid phase could operate. Simply by adding solvent until the last trace of the minor component of the scalemic mixture is washed out would yield the solid enantiopure major component, if we started with an initial scalemic conglomerate.

The case of a racemic compound:

The racemic compound is by itself a crystal phase, $LD_{(s)}$, which can coexist with either the $L_{(s)}$, or the $D_{(s)}$, but not both at the same time. Adding the solvent, we have the three-component or ternary system, $L_{(s)}$ or $D_{(s)}$, $LD_{(s)}$,

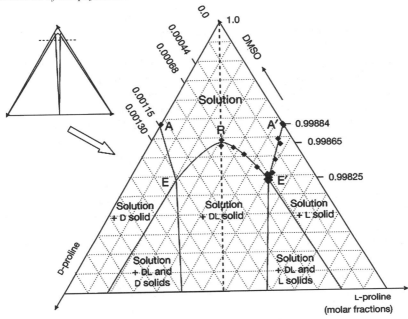

Figure 5.6 Expansion of the apex of the ternary phase diagram of L-proline, D-proline and DMSO (dimethyl sulfoxide) at 25 °C and 1 atm. Saturated solutions exist along the line A–E–R–E′–A′. A and A′ are the solubilities of the enantiopure proline ($L_{(s)}$ or $D_{(s)}$), respectively, while R is the solubility of the racemic compound ($LD_{(s)}$). E and E′ represent the eutectic composition, which is the composition of the solution as long as $LD_{(s)}$, $L_{(s)}$ or $D_{(s)}$, respectively, and saturated solution coexist. The composition of a given point (e.g., E or E′) in the phase diagram can be easily read from the triangular coordinate system (diagram background). For this system, the racemate (R) is only slightly more soluble that the enantiopure compound (A or A′) displaying an eutectic $ee \approx 50\%$, while the proline system in $CHCl_3$ displays the opposite behavior, being characterized by the low solubility of the racemic compound, and therefore high eutectic ee ($ee \approx 99\%$). (Graphic from Reference,[92] with authorization.)

and S (solvent). Gibbs's phase rule can be applied in the same fashion as before, $F + P = C + 2$, and therefore $F = 5 - P$. A two-dimensional plot is obtained by keeping two variables, T and P, fixed. The ternary diagram resembles two symmetric conglomerate diagrams combined by one of the sides of an equilateral triangle (Figure 5.6). As mentioned before, from a given overall composition, i.e., a point inside the triangle, the length of the perpendicular lines to the sides of the equilateral triangle gives the molar fraction of the three components, easily read by using the triangular coordinate. Considering now the relevant case of three phases coexisting in equilibrium, consisting of solid of the racemic compound $LD_{(s)}$, crystals of one enantiomer in excess, $L_{(s)}$ or $D_{(s)}$, and the solution itself, $F = 3 - P = 3 - 3 = 0$; this implies a system with zero degrees of freedom, and again this indicates a fixed composition of the solution phase at the given pressure and temperature. This is the eutectic composition, and the eutectic point is readily visible in the phase diagram (E and E').[92] The composition of the eutectic point may fall anywhere between 0% and 100% *ee*, since it depends on the solubilities of the racemic and enantiopure crystal phases, and these may differ considerably. This is easily understood, since in the eutectic solution the concentration of the minor enantiomer depends entirely on the solubility of the racemic crystal. With independence of the overall *ee* in the initial solid scalemic mixture, its solution will display a fixed composition of eutectic x_L and x_D, and therefore a fixed *ee* %, as long as the three phases coexist.

Determination of the ternary phase system for standard proteinogenic amino acids has been carried out recently.[93] Among the 20 proteinogenic amino acids, all but two crystallize as racemic compounds. Threonine (see also Table 5.2) and arginine form conglomerates and therefore exhibit a 0% *ee* eutectic composition. For some of the remaining amino acids the eutectic *ee* in water is given in Table 5.3. The case of proline is of special significance. Different phases can be obtained under different crystallization conditions, and each phase displays its own physical properties, including solubility. In this case, a $CHCl_3$ molecule is incorporated to the racemic proline unit cell, imparting reduced solubility to the racemate.[93] This type of behavior can be extended in principle to any crystallizing compound. If accepted within the crystal structure, a variety of co-crystallizing additives can extensively alter the properties of the crystallite.

Some of these amino acids provide essentially enantiopure eutectic solutions (*e.g.*, serine, Table 5.3), despite the fact that they might have originated from almost racemic mixtures in the solid state. Recognizing that most chemical reactions occur in solution, the partitioning between different phases constitutes a reliable mechanism for chiral amplification for all asymmetric syntheses, catalytic or not, carried out in solution.

Table 5.3 Eutectic *ee* values for amino acids in water at 25 °C.

Amino acid	ee (%)
Alanine	60
Histidine	94
Isoleucine	52
Leucine	88
Methionine	85
Phenyalanine	83
Proline	50[a]
Serine	99
Threonine	0
Valine	47
Histidine	94

[a]In dimethyl sulfoxide, see Figure 5.6, but 99% *ee* in $CHCl_3$.[94]

5.6.2 Sublimation Properties

The vapor pressure of enantiopure and racemic compounds is equally affected by the nature of the intermolecular forces keeping these condensate phases together. Broadly speaking, this could include both the liquid and solid phases. However, the exceedingly small differences of these diastereomeric interactions in the liquid phase makes the separation of a scalemic mixture (*e.g.*, by distillation) extremely difficult, although not impossible. These differences are about a thousand times larger in the solid phase than in the liquid state. Indeed, the energies involved are perfectly measurable, as shown in Table 5.2. Diastereomeric interactions between enantiomers can also be assessed by comparing the heats of sublimation of the enantiomerically pure sample and the racemic compound. This has been carried out for a limited number of compounds,[95] displaying differences of molar sublimation enthalpies, ΔH_{sublim} (enantiopure) − ΔH_{sublim} (racemic) = ΔH_{mix} [DL$_{(s)}$], of either negative or positive sign for the reasons indicated.[96] The sublimation of enantiomer mixtures can also be illustrated by phase diagrams, resembling those of solid–liquid phases. However, in most cases sublimation phenomena occur under non-equilibrium conditions and may be therefore subject to kinetic effects, in which case the consistency of the phase diagrams is in doubt. Early observations of enantiomeric enrichment by fractional sublimation of an amino acid derivative, in particular of phenylalanine, have been described,[97] this effect being correctly attributed to differences in reticular forces.[98] Indeed, attention was brought to the technique of enhancement of optical activity by fractional sublimation as an efficient alternative to fractional crystallization.[99]

Some recent examples of enantioenrichment of amino acids by sublimation under high vacuum to atmospheric pressure conditions, confirm that this is indeed a practical technique for amplification of scalemic mixtures.[100,101] The occurrence of low *ee* found in amino acid samples from meteorites (Chapter 7, *Outside Earth: Meteorites and Comets*) provides evidence of the presence of scalemic mixtures of amino acids in the solar system. Mechanisms of sublimation and later condensation in the upper atmosphere with *ee* enrichment,[100] followed by precipitation to the Earth, could be related to the meteoritic Late Heavy Bombardment period (LHB or lunar cataclysm), approximately 3800 to 4100 million years ago, in a sequence of events leading to the first enantiomeric imbalance brought to Earth shortly before life appeared.

In spite of the standard deviation from exact racemic, inherent to most racemic mixtures, ($\sigma = \sqrt{N}/2$, where N is the number of molecules in the sample, Section 2.2.1.1, *The Racemic State*), there are no examples of success in the amplification of such a vanishingly small level of *ee* using any of these classical physical techniques. The feasibility of this type of amplification of chirality based on equilibria between phases ultimately rests on the availability of scalemic mixtures. Scalemic mixtures may appear from achiral phases under the effect of a chiral influence. As mentioned in Section 2.3, *Deterministic Theories*, a chiral influence is the key step for a deterministic mechanism to be considered. Once this has occurred, the expected behavior of a scalemic mixture, *i.e.*, fractionation of components between phases, could in principle be the mechanism for further chiral amplification. A completely different scenario corresponds to spontaneous symmetry breaking, which may occur during crystallization; this is discussed in Chapter 6, *Spontaneous Symmetry Breaking*.

5.7 Amplification of Chirality in Serine Octamers

Serine octamers comprise an unusually stable cluster consisting of eight molecules of serine, held together through non-covalent bonds but of uncertain precise structure (Scheme 5.6, *left*).[102] In the first studies,[103] electrospray ionization of an aerosol of serine in methanol–water resulted in a mass spectrum with a prominent ion peak at m/z 841, corresponding to the cationic formula $[\text{Ser}_8 + \text{H}]^+$ (Figure 5.7). Smaller and larger clusters, such as $[\text{Ser}_7 + \text{H}]^+$ and $[\text{Ser}_9 + \text{H}]^+$, are virtually absent in the mass spectrum and eight has therefore been denoted a "magic number" for this kind of association. Serine is rather unique in this respect, the clustering tendency of other aminoacids, although these do exist (*e.g.*, threonine octamers and proline dodecamers, as the most prominent after serine octamers), they are much smaller. A systematic study of clustering caused by sublimation of amino acid crystals shows that the octamer is the most stable cluster, confirming its uniqueness among amino acids, as can be seen in Figure 5.8.[102]

Of special relevance in serine octamers is the strong preference for homochirality exhibited by these clusters, which takes place with a certain degree of chiral amplification, as well as the enantioselectivity displayed in their substitution reactions.

5.7.1 Homochiral Preference and Chiral Amplification

With regard to the first statement above, a strong preference for homochiral clusters is displayed when isotopically labeled racemic serine (consisting of equimolar amounts of 2,3,3-L-$[\text{D}_3]$Ser and D-Ser) is employed. The corresponding homoclusters $[\text{L-}[\text{D}_3]\text{Ser}_8 + \text{H}]^+$ and $[\text{D-Ser}_8 + \text{H}]^+$ are preferred, while mixed clusters of D- and L-serine are significantly under-represented in relation to their statistical weight. The occurrence of these clusters has been also

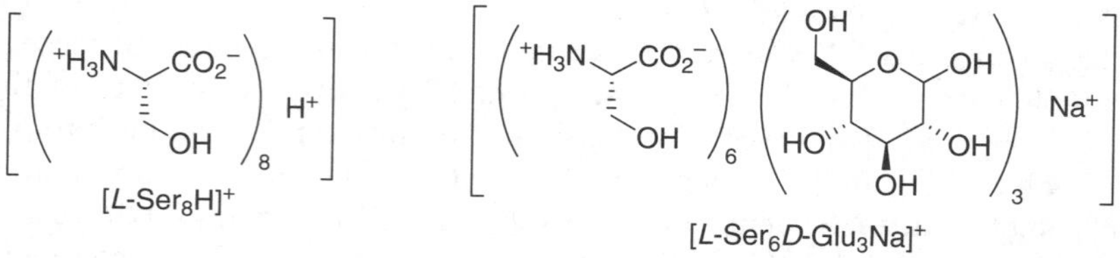

[L-Ser$_8$H]$^+$

[L-Ser$_6$D-Glu$_3$Na]$^+$

Scheme 5.6 The substitution reaction between the L-serine octamer and D-glucose is favored in respect to the diastereomeric L-serine/L-glucose combination.

Mechanisms of Amplification

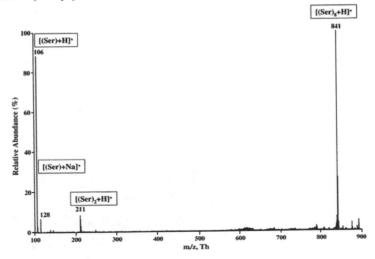

Figure 5.7 Mass spectrum of a 0.01 M L-serine solution prepared in a 1:1 methanol–water mixture and recorded by positive ion electrospray ionization (ESI/MS). (Graphic from Reference 102, with permission.)

established from the negatively charged halide adducts of serine octamers, such as $[Ser_8 + 2Cl]^{-2}$ and $[Ser_8 + 2Br]^{-2}$, in the negative ion electrospray mass spectra of solutions containing serine and the halide, which also displays a homochiral preference.[104] The neutral octamer $[Ser_8]$—undetectable by mass spectroscopy—is still unknown. However, the fact that similar octamers are detected under positive and negative ion mass spectroscopy, both displaying a number of similarities in their chemical properties, suggests that these may originate from the yet undetected neutral octamer $[Ser_8]$.

This preference for homochirality in cluster formation occurs with associated chiral amplification. As determined by a number of mass spectroscopic techniques for a scalemic (or enantioenriched) solution of L-Ser and D-Ser, an enrichment of the *ee* of the serine octamers in relation to the *ee* of the bulk solution is detected (Figure 5.9).[105] Thus, a serine solution of 20% *ee* (Figure 5.9c and 5.9f), produces *ca.* 50% *ee* in the overall octameric cluster composition. In addition, it is obvious from Figure 5.9 that the amounts of minor homochiral octamer (*e.g.*, L_8 in b and d, or D_8 in e and f, Figure 5.9) are clearly underrepresented and do not reflect the *ee* of the bulk solution. This is the case unless the mixture is strictly racemic (Figure 5.9g), in which case, and given the fact that we are not dealing with any mirror-symmetry breaking, the distribution of peaks becomes symmetrical.

5.7.2 Chemistry of Serine Octamers

Substitution reactions in serine clusters have been observed, and these processes take place with a concomitant chiral recognition of the incoming molecule.

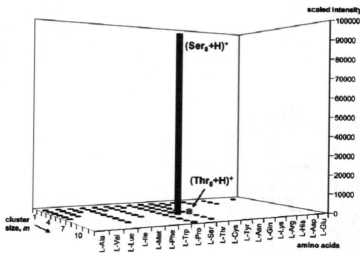

Figure 5.8 Serine forms a homochiral octameric cluster of enhanced stability in comparison with other amino acids. These data were obtained by sublimation from the amino acid crystal. Scaled intensities correspond to $I_m^2/(I_{m-1} \times I_{m+1})$, where I_m is the intensity of the cluster signal of size m. (Graphic from Reference 102, with permission.)

Interestingly, these reactions are favored for the enantiomeric combinations which exist in living matter, namely L-serine with L-amino acids and L-serine with D-sugars. Thus, other amino acids can substitute serine in the octamer, with a strong chiral preference in favor of homochiral cluster formation, especially in the case of threonine and cysteine.[106] Phenylalanine, asparagine, tryptophan and tyrosine can also substitute into the serine octamer; however, the process yields only two such instances and only small chiral effects. In the case of sugars, glucose was incorporated with a magic number cluster corresponding to $[\text{L-Ser}_6\text{D} - \text{Glc}_3\text{Na}]^+$ in the sonic-spray ionization (SSI) mass spectrum of a L-serine/D-glucose mixture in the presence of sodium cations. SSI mass spectra for serine/glucose cluster ions show the preferential formation of the D-glucose/L-serine and L-glucose/D-serine pairs over their L/L and D/D counterparts, using isotopic labeling (Scheme 5.6, *right*).

Intriguingly, serine clusters incorporating other molecules or ions of prebiotic relevance, such as glyceraldehyde or phosphate ions, are also reported. Phosphoric acid incorporates readily into serine clusters by substituting for serine, giving a series of ions such as $[\text{Ser}_{8-n} + (\text{H}_3\text{PO}_4)_n + \text{H}]^+$, ($n = 1-3$). By contrast, sulfuric acid was not incorporated into clusters of serine under analogous conditions. These data may suggest a unique connection between serine and many compounds of fundamental importance in biochemistry.

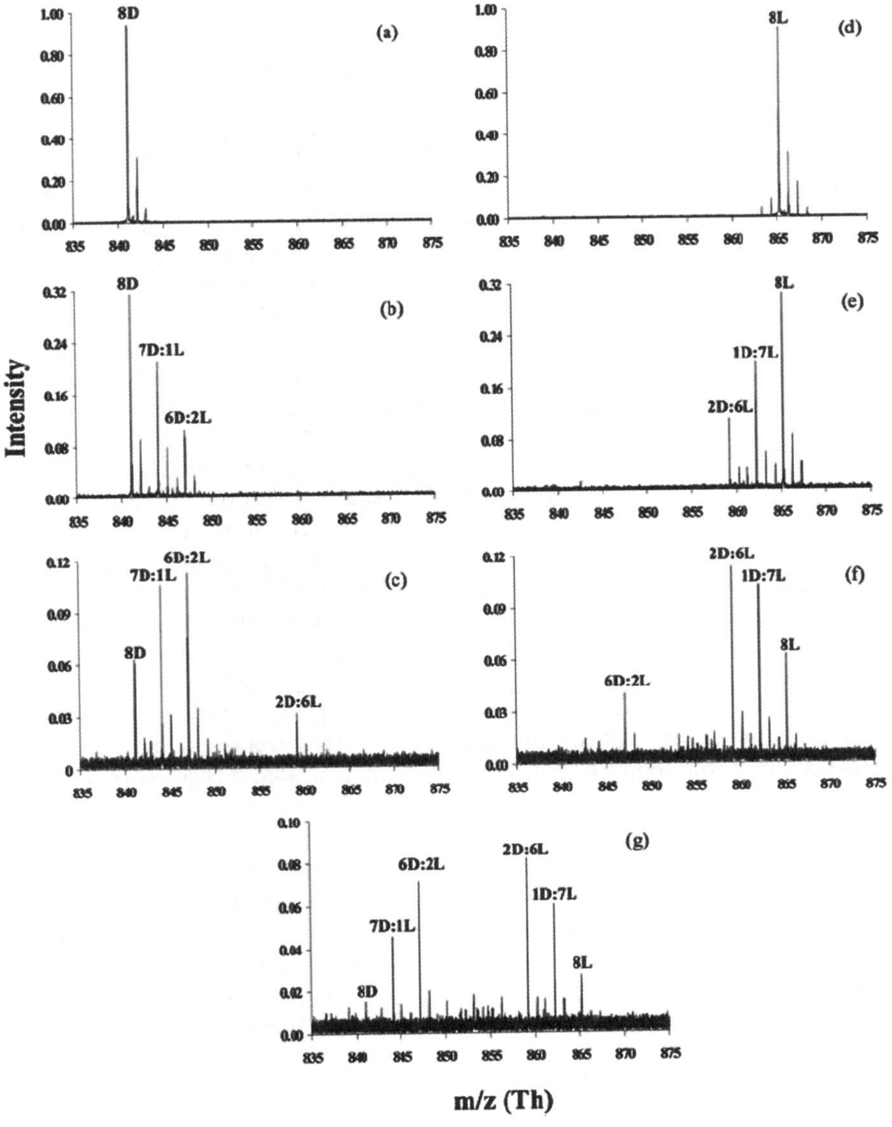

Figure 5.9 Enantiomeric enrichment is evident in the mass spectra of serine mixtures (D-serine, 105 Da; 2,3,3-d_3-L-serine, 108 Da) corresponding to solutions containing different enantiomeric compositions: (a) 100% D, 0% L; (b) 80% D, 20% L; (c) 60% D, 40% L; (d) 0% D, 100% L; (e) 20% D, 80% L; (f) 40% D, 60% L; (g) 50% D, 50% L. A serine solution with small enantiomeric excess yields octamers which are further enriched in the major optical isomer. (Graphic from Reference 106, with permission.)

5.7.3 Sublimation Experiments

Early experiments on serine sublimation, as well as studies with other amino acids, have been reported.[107] Intense ion formation was observed between 200 °C and 300 °C, with more than 99% of the observed ion present corresponding to protonated serine octamer ions [Ser$_8$ + H]$^+$, as a single species, not a statistical distribution of clusters. In the case of racemic serine crystals no octamer was present, but intense monomer and dimer signals were detected; by contrast in this case a mixture of D- and L-serine gave a signal comparable in intensity to pure L-serine. The observed difference between the mixture of L- and D-serine and DL-serine crystals suggests the direct transfer of octameric units from the condensed to the gas phase, rather than re-assembly in the gas phase either before or after the ionization event to give the observed octameric species. A comparison of the characteristics of the serine octamers obtained by sonic/electrospray and thermal methods reveals that the main features displayed by [Ser$_8$ + H]$^+$ are similar, including preference for homochirality and chiral transmission to other amino acids.[108] (See also Section 5.6, *Amplification of Scalemic Compounds: Eutectic Mixtures*, and Section 5.6.2, *Sublimation Properties* for related studies on the sublimation of other amino acids.)

The remarkable properties displayed by the serine octamer have been studied by detection of the cluster in the gas phase, mainly in the cationic form [Ser$_8$ + H]$^+$ and related charged adducts. Unfortunately, the neutral octamer of serine, [Ser$_8$], which in water would be of the highest relevance in prebiotic processes such as amplification of chirality and chirality transfer to other molecules, remains elusive. For the moment, attempts to detect serine octamers from aqueous serine solution have not been successful. Infrared measurements, NMR chemical shifts, and particularly the determination of diffusion coefficients at different pH or concentration, do not show evidence of octameric aggregation.[109]

In anticipation of further studies in the aqueous phase, it can be speculated that serine clusters were involved in chiral accumulation and in transmission of chirality to other amino acids and prebiotic molecules. Serine might have self-enriched into its homochiral forms in the course of forming the octamer in concentrated aqueous solution. This might then have served as a site for essential prebiotic reactions, which may include stereoselective aldol condensations of C$_3$ sugars (glyceraldehyde), uptake of phosphoric acid in a form suited to controlled phosphorylation, or metal-ion binding and asymmetric catalysis. Whether these facts took place in a prebiotic scenario or not has yet to be established.

References

1. F. C. Frank, *Biochem. Biophys. Acta.*, 1953, **11**, 459–463.
2. D. P. Kondepudi and I. Prigonine, *Modern Thermodynamics: From Heat Engines to Dissipative Structures*, John Wiley and Son Ltd., Chichester, 1998, 431–438.

3. D. G. Blackmond, *Proc. Natl. Acad. Sci. USA*, 2004, **101**, 5732–5736.
4. (a) J. M. Ribó, J. Crusats, F. Sagués, J. Claret and R. Rubires, *Science*, 2001, **292**, 2063–2066; (b) R. Rubires, J-A. Farrera and J. M. Ribó, *Chem. Eur. J.*, 2001, **7**, 436–446.
5. (a) X. Qiu, J. Ruiz-Garcia, K. J. Stine, C. M. Knobler and J. V. Selinger, *Phys. Rev. Lett.*, 1991, **67**, 703–706; (b) D. K. Schwartz, R. Viswanathan and J. A. N. Zasadzinski, *Phys. Rev. Lett.*, 1993, **70**, 1267–1270; (c) H. V. Berlepsch, C. Böttcher, A. Quart, S. Dähne and S. Kirstein, *J. Phys. Chem. B*, 2000, **104**, 5255–5262.
6. R. Plasson, D. K. Kondepudi, H. Bersini, A. Commeyras and K. Asakura, *Chirality*, 2007, **19**, 589–600.
7. D. K. Kondepudi and G. W. Nelson, *Phys. Rev. Lett.*, 1983, **50**, 1023–1026.
8. Y. Saito and H. Hyuga, *J. Phys. Chem. Jpn.*, 2004, **73**, 1685–1688.
9. (a) M. Calvin, *Chemical Evolution*, Oxford University Press, Oxford, 1969; (b) K. Asakura, T. Soga, T. Uchida, S. Osanai and D. K. Kondepudi, *Chirality*, 2002, **14**, 85–89.
10. Y. Saito and H. Hyuga, *J. Phys. Soc. Jpn.*, 2004, **73**, 33–35.
11. R. Plasson, H. Bersini and A. Commeyras, *Proc. Natl. Acad. Sci. USA*, 2004, **101**, 16733–16738.
12. (a) P. D. Bartlett and R. H. Jones, *J. Am. Chem. Soc.*, 1957, **79**, 2153–2159; (b) R. D. Lundberg and P. Doty, *J. Am. Chem. Soc.*, 1956, **78**, 4810–4812.
13. (a) K. Iwamoto, *Phys. Chem. Chem. Phys.*, 2003, **5**, 3616–3621; (b) J. R. Islas, D. Lavabre, J. M. Grevy, R. H. Lamoneda, H. R. Cabrera, J. C. Micheau and T. Buhse, *Proc. Natl. Acad. Sci. USA*, 2005, **102**, 13743–13748.
14. P. G. H. Sandars, *Orig. Life Evol. Biosph.*, 2003, **33**, 575–587.
15. Y. Saito and H. Hyuga, *J. Phys. Soc. Jpn.*, 2005, **74**, 1629–1635.
16. (a) Y. Saito and H. Hyuga, *J. Phys. Chem. Jpn.*, 2004, **73**, 1685–1688; (b) Y. Saito and H. Hyuga, *J. Phys. Soc. Jpn.*, 2005, **74**, 535–537.
17. D. K. Kondepudi and G. W. Nelson, *Nature*, 1985, **314**, 438–441.
18. D. K. Kondepudi and K. Asakura, *Acc. Chem. Res.*, 2001, **34**, 946–954.
19. (a) J. M. Ribó, J. Crusats, F. Sagués, J. Claret and R. Rubires, *Science*, 2001, **292**, 2063–2066; (b) R. Rubires, J-A. Farrera and J. M. Ribó, *Chem. Eur. J.*, 2001, **7**, 436–446.
20. K. Soai, S. Niwa and H. Hori, *J. Chem. Soc., Chem. Commun.*, 1990, **982–983**.
21. T. Shibata, S. Yonekubo and K. Soai, *Angew. Chem., Int. Ed.*, 1999, **38**, 659–661.
22. (a) I. Sato, D. Omiya, K. Tsukiyama, Y. Ogi and K. Soai, *Tetrahedron: Asymmetry*, 2001, **12**, 1965–1969; (b) I. Sato, D. Omiya, H. Igarashi, K. Kato, Y. Ogi, K. Tsukiyama and K. Soai, *Tetrahedron: Asymmetry*, 2003, **14**, 975–979.
23. (a) T. Buhse, *Tetrahedron: Asymmetry*, 2003, **14**, 1055–1061; (b) I. D. Gridnev, J. M. Serafimov and J. M. Brown, *Angew. Chem., Int. Ed.*, 2004,

43, 4884–4887; (c) D. G. Blackmond, *Proc. Natl. Acad. Sci. U.S.A.*, 2004, **101**, 5732–5736.
24. Y. Saito and H. Hyuga, *J. Phys. Soc. Jpn.*, 2004, **73**, 33–35.
25. D. G. Blackmond, *Tetrahedron: Asymmetry*, 2006, **17**, 584–589.
26. I. Sato, H. Urabe, S. Ishiguro, T. Shibata and K. Soai, *Angew. Chem., Int. Ed.*, 2003, **42**, 315–317.
27. Personal communication, Professor K. Soai.
28. K. Soai, I. Sato, T. Shibata, S. Komiya, M. Hayashi, Y. Matsueda, H. Imamura, T. Hayase, H. Morioka, H. Tabira, J. Yamamoto and Y. Kowata, *Tetrahedron: Asymmetry*, 2003, **14**, 185–188.
29. D. A. Singleton and L. K. Vo, *J. Am. Chem. Soc.*, 2002, **124**, 10010–10011.
30. T. Kawasaki, K. Suzuki, M. Shimizu, K. Ishikawa and K. Soai, *Chirality*, 2006, **18**, 479–482.
31. K. Soai and I. Sato, *Chirality*, 2002, **14**, 548–554.
32. I. Sato, Y. Matsueda, K. Kadowaki, S. Yonekubo, T. Shibata and K. Soai, *Helv. Chim. Acta.*, 2002, **85**, 3383–3387.
33. K. Soai and T. Kawasaki, *Chirality*, 2006, **18**, 469–478.
34. T. Kawasaki, H. Tanaka, T. Tsutsumi, T. Kasahara, I. Sato and K. Soai, *J. Am. Chem. Soc.*, 2006, **128**, 6032–6033.
35. (a) K. Soai, T. Shibata and I. Sato, *Acc. Chem. Res.*, 2000, **33**, 382–390; (b) I. Sato, R. Sugie, Y. Matsueda, Y. Furumura and K. Soai, *Angew. Chem., Int. Ed.*, 2004, **43**, 4490–4492.
36. K. Soai, T. Kawasaki and I. Sato, in *The Chemistry of Organozinc Compounds, Part 2*, ed. Z. Rappoport and I. Marek, Wiley, Chichester, 2006, p. 580.
37. T. Kawasaki, M. Sato, S. Ishiguro, T. Saito, Y. Morishita, I. Sato, H. Nishino, Y. Inoue and K. Soai, *J. Am. Chem. Soc.*, 2005, **127**, 3274–3275.
38. K. Soai, K. Osanai, K. Kadowaki, S. Yonekubo, T. Shibata and I. Sato, *J. Am. Chem. Soc.*, 1999, **121**, 11235–11236.
39. I. Sato, K. Kadowaki and K. Soai, *Angew. Chem., Int. Ed.*, 2000, **39**, 1510–1512.
40. I. Sato, K. Kadowaki, Y. Ohgo and K. Soai, *J. Mol. Catal. A*, 2004, **216**, 209–214.
41. I. Sato, K. Kadowaki, H. Urabe, J. H. Jung, Y. Ono, S. Shinkai and K. Soai, *Tetrahedron Lett.*, 2003, **44**, 721–724.
42. T. Kawasaki, K. Ishikawa, H. Sekibata, I. Sato and K. Soai, *Tetrahedron Lett.*, 2004, **45**, 7939–7941.
43. I. Sato, M. Shimizu, T. Kawasaki and K. Soai, *Bull. Chem. Soc. Jpn.*, 2004, **77**, 1587–1588.
44. T. Kawasaki, K. Jo, H. Igarashi, I. Sato, M. Nagano, H. Koshima and K. Soai, *Angew. Chem., Int. Ed.*, 2005, **44**, 2774-2777.
45. (a) T. Inoue and L.E. Orgel, *J. Molec. Biol.*, 1982, **162**, 204-217; (b) G. von Kiedrowski, *Angew. Chem., Int. Ed. Engl.*, 1986, **25**, 932–935; (c) E. A. Wintner, M. M. Conn and J. Rebek, Jr., *Acc. Chem. Res.*, 1994, **27**, 198–203.

46. D. Sievers and G. von Kiedrowski, *Nature*, 1994, **369**, 221–224.
47. G. F. Joyce, G. M. Visser, C. A. van Boeckel, J. H. van Boom, L. E. Orgel and J. van Westrenen, *Nature*, 1984, **310**, 602–604.
48. J. G. Schmidt, P. E. Nielsen and L. E. Orgel, *J. Am. Chem. Soc.*, 1997, **119**, 1494–1495.
49. H. L. Lee, J. R. Granja, J. A. Martinez, K. Severin and M. R. Ghadiri, *Nature*, 1996, **382**, 525–528.
50. (a) A. Saghatelian, Y. Yokobayashi, K. Soltani and M. R. Ghadiri, *Nature*, 2001, **409**, 797–801; (b) K. Severin, D. H. Lee, J. A. Martinez, M. Vieth and M. R. Ghadiri, *Angew. Chem., Int. Ed.*, 1998, **37**, 126–128.
51. (a) S. Yao, I. Ghosh, R. Zutshi and J. Chmielewski, *Nature*, 1998, **396**, 447–450; (b) X. Li and J. Chmielewski, *Org. Biomol. Chem.*, 2003, **1**, 901–904.
52. L. Leman, L. Orgel and M. R. Ghadiri, *Science*, 2004, **306**, 283–286.
53. (a) M. G. Schwendinger and B. M. Rode, *Anal. Sci.*, 1989, **5**, 411–414; (b) B. M. Rode, *Peptides*, 1999, **20**, 773–786.
54. C. Puchot, O. Samuel, E. Dunach, S. Zhao, C. Agami and H. B. Kagan, *J. Am. Chem. Soc.*, 1986, **108**, 2353–2357.
55. (a) H. B. Kagan, *Adv. Synth. Catal.*, 2001, **343**, 227–233; (b) H. B. Kagan, *Synlett.*, 2001, 888–899; (c) H. B. Kagan, C. Girard, D. Guillaneux, D. Rainford, O. Samuel, S. Y. Zhang and S. H. Zhao, *Acta Chem. Scand.*, 1996, **50**, 345–352.
56. D. Guillaneux, S. H. Zhao, O. Samuel, D. Rainford and H. B. Kagan, *J. Am. Chem. Soc.*, 1994, **116**, 9430–9439.
57. C. Girard and H. B. Kagan, *Angew. Chem., Int. Ed.*, 1998, **37**, 2923–2959.
58. T. Akaike, Y. Aogaki and S. Inoue, *Biopolymers*, 1975, **14**, 2577–2583.
59. T. H. Hitz and P. L. Luisi, *Orig. Life Evol. Biosph.*, 2004, **34**, 93–110.
60. T. Hitz and P. L. Luisi, *Helv. Chim. Acta*, 2003, **86**, 1423–1434.
61. (a) T. Tsuruta, S. Inoue and K. Matsuura, *Biopolymers*, 1967, **5**, 313–319; (b) W. A. Bonner, N. E. Blair and F. M. Dirbas, *Orig. Life*, 1981, **11**, 119–134.
62. H. Yue and H. Wang, *Measurement and Control*, 2003, **36**, 209–215.
63. M. M. Green, N. C. Peterson, T. Sato, A. Teramoto, R. Cook and S. Lifson, *Science*, 1995, **268**, 1860–1866.
64. M. M. Green, M. P. Reidy, R. D. Johnson, G. Darling, D. J. O'Leary and G. Willson, *J. Am. Chem. Soc.*, 1989, **111**, 6452–6454.
65. (a) M. M. Green, B. A. Garetz, B. Munoz, H. Chang, S. Hoke and R. G. Cooks, *J. Am. Chem. Soc.*, 1995, **117**, 4181–4182; (b) J. van Gestel, *Macromolecules*, 2007, **37**, 3894–3898.
66. M. M. Green, J.-W. Park, T. Sato, A. Teramoto, S. Lifson, R. L. B. Selinger and J. V. Selinger, *Angew.Chem., Int. Ed.*, 1999, **38**, 3138–3154.
67. (a) A. Teramoto, *Prog. Polym. Sci.*, 2001, **26**, 667–720; (b) F. Takei, K. Yanai, K. Onitsuka and S. Takahashi, *Chem. Eur. J.*, 2000, **6**, 983–993.
68. (a) *Chirality in Liquid Crystals*, ed. H-S. Kitzerow and C. Bahr, Springer, Secaucus, N. J., 2001; (b) A. Jakli, G. G. Nair, C. K. Lee, R. Sun and L. C. Chien, *Phys. Rev. E*, 2001, **63**, 061710/5.

69. E. Yashima, K. Maeda and T. Nishimura, *Chem. Eur. J.*, 2004, **10**, 42–51.
70. R. Eelkema and B. L. Feringa, *Org. Biomol. Chem.*, 2006, **4**, 3729–3745.
71. A. R. A. Palmans and E. W. Meijer, *Angew. Chem., Int. Ed.*, 2007, **46**, 8948–8968.
72. Y. Yamagata, *J. Theor. Biol.*, 1966, **11**, 495–498.
73. Y. Yamagata, H. Sakihama and K. Nakano, *Orig. Life*, 1980, **10**, 349–355.
74. F. Faglioni, A. Passalacqua and P. Lazzeretti, *Orig. Life Evol. Biosph.*, 2005, **35**, 461–475.
75. A. J. MacDermott, G. E. Tranter and S. J. Trainor, *Chem. Phys. Lett.*, 1992, **194**, 152–156.
76. D. K. Kondepudi, R. J. Kaufman and N. Singh, *Science*, 1990, **250**, 975–977.
77. K. Evgenii and T. Wolfram, *Orig. Life Evol. Biosph.*, 2000, **30**, 431–434.
78. G. This ee corresponds to a crystal size of ca. 0.1 mm: E. Tranter, *Nature*, 1985, **318**, 172–173.
79. W. A. Bonner, *Orig. Life Evol. Biosphere*, 1999, **29**, 615–623.
80. L. Keszthelyi, *Orig. Life Evol. Biosphere*, 2001, **31**, 249–256.
81. (a) A. Salam, *J. Mol. Evol.*, 1991, **33**, 105–113; (b) A. Salam, in *Chemical Evolution: Origin of Life*, ed. C. Ponnamperuma and J. Chela-Flores, A. Deepak Publ., Hampton, VA, 1993, p. 101.
82. A. Figureau, E. Duval and A. Boukenter, *Orig. Life Evol. Biosphere*, 1995, **25**, 211–217.
83. W. Wang, F. Yi, Y. Ni, Z. Zhao, X. Jin and Y. Tang, *J. Biol. Phys.*, 2000, **26**, 51–65.
84. W.-Q. Wang, W. Min, L. Zhi, L.-Y. Wang, L. Chen and F. Deng, *Biophys. Chem.*, 2003, **103**, 289–298.
85. R. Sullivan, M. Pyda, J. Pak, B. Wunderlich, J. R. Thompson, R. Pagni, H. Pan, C. Barnes, P. Schwerdtfeger and R. Compton, *J. Phys. Chem. A*, 2003, **107**, 6674–6680.
86. W.-Q. Wang, Y.-N. Liu and Y. Gong, *Wuli Huaxue Xuebao*, 2004, **20**, 1345–1351.
87. D. G. Blackmond and M. Klussmann, *Chem. Commun.*, 2007, 3990–3996.
88. E. L. Eliel, S. H. Wilen and L. N. Mander, *Stereochemistry of Organic Compounds*, John Wiley & Sons, New York, 1994, p. 159.
89. H. D. Flack, *Helv. Chim. Acta*, 2003, **86**, 905–921.
90. M. Matsumoto and K. Amaya, *Bull. Chem. Soc. Jpn.*, 1980, **53**, 3510–3512.
91. S. Takagi, R. Fujishiro and K. Amaya, *Chem. Commun.*, 1968, 480.
92. M. Klussmann, H. Iwamura, S. P. Mathew, D. H. Wells Jr, U. Pandya, A. Armstrong and D. G. Blackmond, *Nature*, 2006, **441**, 621–623.
93. M. Klussmann, A. J. P. White, A. Armstrong and D. G. Blackmond, *Angew. Chem., Int. Ed.*, 2006, **45**, 7985–7989.
94. Y. Hayashi, M. Matsuzawa, J. Yamaguchi, S. Yonehara, Y. Matsumoto, M. Shoji, D. Hashizume and H. Koshino, *Angew. Chem., Int. Ed.*, 2006, **45**, 4593–4597.

95. J. S. Chickos and D. G. Hesse, *Structural Chem.*, 1991, **2**, 33–40.
96. V. A. Soloshonok, H. Ueki, M. Yasumoto, S. Mekala, J. S. Hirschi and D. A. Singleton, *J. Am. Chem. Soc.*, 2007, **129**, 12112–12113.
97. G. Pracejus, *Justus Liebigs Ann. Chem.*, 1959, **622**, 10–22.
98. H. Kwart and D. P. Hoster, *J. Org. Chem.*, 1967, **32**, 1867–1871.
99. D. L. Garin, D. J. Cooke Greco and L. Kelley, *J. Org. Chem.*, 1977, **42**, 1249–1251.
100. S. P. Fletcher, R. B. C. Jagt and B. L. Feringa, *Chem. Commun.*, 2007, 2578–2580.
101. R. H. Perry, C. Wu, M. Nefliu and R. G. Cooks, *Chem. Commun.*, 2007, 1071–1073.
102. S. C. Nanita and R. G. Cooks, *Angew. Chem., Int. Ed.*, 2006, **45**, 554–569.
103. (a) R. G. Cooks, D. Zhang, K. J. Koch, F. C. Gozzo and M. N. Eberlin, *Anal. Chem.*, 2001, **73**, 3646–3655; (b) R. Hodyss, R. R. Julian and J. L. Beauchamp, *Chirality*, 2001, **13**, 703–706.
104. S. C. Nanita and R. G. Cooks, *J. Phys. Chem. B*, 2005, **109**, 4748–4753.
105. S. C. Nanita, Z. Takats, R. G. Cooks, S. Myung and D. E. Clemmer, *J. Am. Soc. Mass Spectrom.*, 2004, **15**, 1360–1365.
106. Z. Takats, S. C. Nanita, G. Schlosser, K. Vekey and R. G. Cooks, *Anal. Chem.*, 2003, **75**, 6147–6154.
107. (a) Z. Takats and R. G. Cooks, *Chem. Commun.*, 2004, 444–445; (b) V. A. Yaylayan, A. Keyhani and A. Wnorowski, *J. Agric. Food Chem.*, 2000, **48**, 636–641.
108. P. Yang, R. Xu, S. C. Nanita and R. G. Cooks, *J. Am. Chem. Soc.*, 2006, **128**, 17074–17086.
109. S. Vandenbussche, G. Vandenbussche, J. Reisse and K. Bartik, *Eur. J. Org. Chem.*, 2006, 3069–3073.

CHAPTER 6
Spontaneous Symmetry Breaking

6.1 Introduction

The spontaneous generation of chirality is not an intuitive concept. Might it be possible to generate chirality—for instance, during a reaction—using neither chiral reagents nor other sources of asymmetry? During recent decades, a small number of experiments have demonstrated that this is indeed possible.[1]

All the scenarios in which spontaneous symmetry breaking has occurred share certain features in common:

- Firstly, they are all systems far from equilibrium. Under homogeneous thermodynamic equilibrium there can be no enantiomeric excess. In such a case, if a molecule is chiral, *i.e.*, it has two possible states of equal energy, the entropy of mixing assures that the lowest possible free energy state corresponds to a 1 : 1 mixture of each enantiomer. This result is obtained from the classical definition of entropy of mixing, $\Delta S_{mix} = -nR\Sigma_i x_i \ln x_i$ (1), where n is the overall number of moles of the mixture, composed of i components, each one of x_i mole fraction. For the mixture of two components, *i.e.*, both enantiomers, this function has a maximum at $x_i = 0.5$, giving the well known expression $\Delta S_{mix} = R \ln 2$, per mol of racemic mixture. It must be stressed that this expression is derived from calculation of the expansion of ideal gases followed by mixing, and inherently implies particles indistinguishable within the mixture. A similar conclusion is arrived at by statistical thermodynamic reasoning, since in this situation the number of possible microstates is maximized. Thus from the Boltzmann entropy formula, $\Delta S_{mix} = k \ln W_{mix}$, where $W_{mix} = N!/\Pi_i N_i!$, is the increase in the number of microstates in the mixing process of i distinguishable components, and $N = \Sigma_i N_i$ represents the overall number of

---

The Origin of Chirality in the Molecules of Life
Albert Guijarro and Miguel Yus
© Albert Guijarro and Miguel Yus, 2009
Published by the Royal Society of Chemistry, www.rsc.org

Spontaneous Symmetry Breaking

molecules, an expression for the entropy of mixing (1) is equally obtained by applying the Stirling approximation, $\ln N! \approx N \ln N - N$, as well as observing that $R = k N_A$.[2] It is patent in this example that the mixing molecules must be indistinguishable within the mixture, and this is consistent with a homogeneous solution. The situation may be well different in scenarios in which different phases emerge. In the crystal phase molecules are no longer indistinguishable, so the entropy of the system is not described adequately by the above equations. Although the number of microstates seems to be maximized if crystals of both handedness are present, the final free energy balance may not exclusively be driven by entropic considerations.

- Secondly, they appear as a manifestation of some kind of chiral autocatalysis, which we have seen is capable of amplifying an initial chiral bias. But where does this initial chiral bias come from? Since we have prescribed the absence of any chiral influence in the reaction, the answer is not obvious.
- This leads to the third characteristic of these processes: the overall trend of the system obeys spontaneous stochastic fluctuations of its constituent molecules (see for instance Section 2.2.1.1, *The Racemic State*). Under appropriate conditions—seldom fulfilled—these fluctuations can be amplified and are eventually manifested as macromolecular chirality.

Due to its intrinsic nature, stochastic fluctuation implies that an experiment follows a statistical behavior. The chances of chiral fluctuation in one particular sense (*e.g.*, *plus*) is necessarily 50%, for reasons of symmetry, and over a large series of experiments a further 50% of these can be expected to produce the opposite chirality (*i.e.*, *minus*); it is thus not possible to anticipate the chiral outcome of any given experiment. From this we can argue that spontaneous symmetry breaking is only breaking the symmetry if we do not take into account the overall picture, which includes stochastic behavior of the system in agreement with what is known in statistics as the law of large numbers.[3] Over a large number of experiments symmetry is thus again restored. In addition to the three characteristics mentioned, phase transition is a further recurrent feature of these systems, of which the relevance has still to be determined. The appearance of new phases is in principle not required by the Frank model, which is built on normal homogeneous kinetics. However, the association of this phenomenon with symmetry breaking scenarios is generally observed—although not always in the form of true phases, but also as mesophases, micelles, etc.

We have previously discussed the case of asymmetric autocatalysis (Section 5.2.4, *Asymmetric Autocatalytic Reactions*), and spontaneous symmetry breaking is discussed in the section which now follows (Section 6.2, *Spontaneous Symmetry Breaking in Crystallization*). Both of these are considered mechanistically in Chapter 5, *Mechanisms of Amplification*. As mentioned previously, with the exception of Soai's addition of diisopropylzinc to 2-alkynylpyrimidine-5-carbaldehyde in diethyl ether (Section 5.2.4., *q. v.*),[4] no chemical reaction has ever displayed this behavior.

The aggregation of mesophases from homogeneous solutions of achiral compounds is another scenario of symmetry breaking (Section 4.4, *Fluid Dynamics: Vortex Motion*), as well as certain aspects of two-dimensional (2D) chemistry in α-glycine crystals at the air–water interface (Section 8.2.3, *Organic Crystals: Glycine*). These have all been described relatively recently. On the other hand, the primary and most studied scenario of spontaneous symmetry breaking is as a physical process, namely, during the crystallization of conglomerates.

6.2 Spontaneous Symmetry Breaking in Crystallization

The archetypal example of spontaneous symmetry breaking is the physical process involving the emergence of a new phase—the crystallization of conglomerates (of structurally enantiomorphic crystals) from supersaturated solutions or supercooled melts.[5] Until recently, these have been the only known examples of spontaneous symmetry breaking. The crystallization of sodium chlorate, an achiral ionic compound (point group ClO_3^-: C_{3v}; Na^+: spherical) crystallizing in the chiral $P2_13$ space group of the cubic system, is a representative case.[6] Sodium chlorate rarely shows crystals with hemihedral faces allowing visual identification of enantiomers, in the manner used by Pasteur. Instead, the two chiral forms of sodium chlorate crystals are birefringent and can be separated by means of the interference colors produced under the analyzer using polarized light (Figure 6.1).

In the absence of any perturbation, crystallization of $NaClO_3$ from water yields a random distribution of (+) and (−) crystals in a ratio close to 1 : 1 (Figure 6.2a), represented in Figure 6.2b as a monomodal distribution of crystal enantiomeric excess (*cee*). If the solution is stirred during crystallization, however, a given experiment yields mostly either (+) or (−) crystals, with no correlation between the direction of stirring and the chirality of the crystals (Figure 6.2c). If sufficient crystallizations are carried out under stirring, an overall random distribution of (+) and (−) crystals is obtained, displaying a bimodal probability distribution (Figure 6.2d), which is the signature of spontaneous symmetry breaking.

This phenomenon is shared by other compounds, among them sodium bromate and 4,4′-dimethyl chalcone.[7] Being achiral molecules which crystallize in a chiral space group, under appropriate crystallization conditions these also display a bimodal statistical distribution of (+) and (−) crystals. A similar effect occurs in the crystallization of chiral 1,1′-binaphthyl from the molten state at 150 °C.[8] Binaphthyl crystallizes at 158 °C as a conglomerate of (+) and (−) atropo-enantiomers, in which not only the crystal but also the molecule itself is chiral. However, in the molten state the enantiomers interconvert rapidly ($t_{1/2} < 1$ s), so the mother liquor is always racemic. The fact that the entire crystallite can be chiral (*ee* > 80%) is in agreement with rupture of symmetry. *N*-aryl-2(1*H*)-pyrimidinones, displaying axial chirality and low racemization barriers, behave similarly.[9] This is not the lowest energy state of the system, which corresponds either to a racemic crystal (racemate) or to a

Spontaneous Symmetry Breaking

Figure 6.1 Crystals of $NaClO_3$ under polarized light. The levo $(-)$ and dextro $(+)$ crystals can easily be identified using a pair of polarizers. Crystal color changes occur on rotating the analyzer clockwise (*left*) or anticlockwise (*right*), depending on the handedness of the crystal. (Pictures courtesy of C. Viedma, Reference 11).

racemic mixture of crystals (conglomerate). Again, we are far from equilibrium, under conditions which involve the emergence of new phases.

In connection with the Frank (or Frank–Kondepudi) model of autocatalysis seen earlier, kinetic models emphasize the importance of secondary nucleation under supersaturated conditions.[5b] In recent work this connection has been supported using simulations of the dynamics of these symmetry breaking systems. These studies arrived at the conclusion that secondary nucleation is a nonlinear autocatalytic phenomenon, which would explain these observations.[10] Nonetheless, the behavior of closed systems far from equilibrium is often poorly understood and is hard to predict, as mentioned in Section 5.2.2, *Theoretical Models Derived from Frank's Original Model*. Operating under different crystallization conditions—in particular a higher degree of supersaturation—almost 100% of crystals of $NaClO_3$ of identical handedness originated from an apparently massive primary nucleation process, as witnessed by videotaping.[11]

By using continuous mechanical crushing using glass balls of the crystals obtained in the classical experiment of crystallization of $NaClO_3$, the conditions required in a recycled Frank model (Section 5.2.1, *The Frank Model*) can

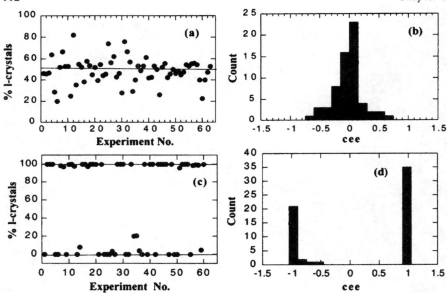

Figure 6.2 Spontaneous chiral symmetry breaking in the crystallization of $NaClO_3$. (a) represents the percentage of (−)-crystals obtained during 63 unstirred crystallizations, and (b) is the corresponding histogram of crystal enantiomeric excess, cee, [cee = $(N_{(-)} - N_{(+)})/(N_{(-)} + N_{(+)})$, where N = number of individual crystals], showing a monomodal probability distribution. (c) represents the percentage of (−)-crystals after 60 stirred crystallizations, with histogram (d) showing a bimodal distribution, associated with stochastic behavior and spontaneous symmetry breaking. (From Reference 5b, with permission.)

be reproduced.[12] The system becomes a closed, non-isolated system which is not under thermodynamic equilibrium, supported by evaluation of the chemical potentials involved. The chemical potential of sodium chlorate in the different phases, μ_{NaClO3}, spontaneously decreases on passing from the supersaturated solution, to small crystals and then to large crystals, while the activation energy for recycling in the opposite direction is mechanically provided in the system (Section 5.2.2, *Theoretical Models Derived from Frank's Original Model*, and Figure 5.2c). The crushing ensures continuous dissolution and crystallization of all the crystals, and therefore no irreversible accumulation of unwanted minor enantiomers occurs and a stationary Frank regime of complete chiral purity is eventually obtained.

This process plays the role of reversibility in open chemical systems and has been proposed for theoretical models, carrying the title "correction mechanisms".[13] It is remarkable that, induced by nonlinear autocatalysis in a closed system impelled solely by mechanical recycling, a perfect homochiral state is accomplished (Figure 6.3). The expression "Ostwald ripening" has been appropriately chosen for the more general treatment described above, based on chemical potentials.[14,15] During crystallization this is a process in which large

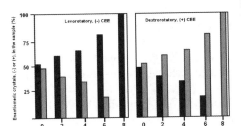
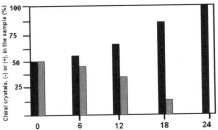

Figure 6.3 A stationary perfect homochiral state is obtained in the crushing–crystallization experiment. Complete chiral amplification is observed (*left*) in solutions with a 5% initial L-*cee* or D-*cee*, increasing to 100% L-*cee* or D-*cee* in 8 h at 600 rpm. Starting with an initial racemic conglomerate of D and L crystals (*right*), a spontaneus symmetry breaking experiment is achieved: 100% *cee* is observed after 24 h at 600 rpm, with random handedness (*cee*: crystal enantiomeric excess). (From Reference 12, with permission.)

crystals grow at the expense of smaller ones, which in turn shrink and ultimately disappear.[16] This spontaneous process is driven by minimization of the number of molecules on the surface of the crystal. These are molecules of high chemical potential in thermodynamic equilibrium with the saturated solution, μ_{NaClO_3} (surface) = μ_{NaClO_3} (solution), but at an energetic disadvantage compared with the molecules inside the reticular structure, μ_{NaClO_3} (surface) > μ_{NaClO_3} (crystal body). Through the continuous medium, a free energy-driven mass transport process will eventually enlarge the larger crystals at the expense of the smaller ones, which have a higher specific surface. The driving force of this mass transport process is inherently small, but is greatly enhanced by extensive ball-milling. It should be noted that the surface area of a crystal in course of being crushed increases, following an exponential function. This can be appreciated from the isomorphic dissection of a cube. If a cubic crystal of edge length l is dissected into eight identical smaller crystals (by the coordinate planes), and each small cubic crystal is dissected again, and then again n times, the resulting overall surface, S, is given by the exponential expression, $S = 6l \times 2^n$.

Of particular importance, this magnification of chirality is the macroscopic manifestation of individual microscopic stochastic events, and is a very rare phenomenon in sciences such as chemistry which deal with very large numbers of molecules and follow well ordered associated statistics. Its extension to other systems more closely related to the prebiotic environment have been proposed,[14] and experiments including a step of chemical racemization in the homogeneous phase reported.[17]

References

1. M. Avalos, R. Babiano, P. Cintas, J. L. Jimenez and J. C. Palacios, *Orig. Life Evol. Biosph.*, 2004, **34**, 391–405.

2. D. P. Kondepudi and I. Prigonine, *Modern Thermodynamics: From Heat Engines to Dissipative Structures*, John Wiley & Son Ltd., Chichester, 1998, 155–157.
3. J. Renze and E. W. Weisstein, *Law of Large Numbers, from MathWorld—A Wolfram Web Resource*, http://mathworld.wolfram.com/LawofLarge Numbers.html.
4. K. Soai, I. Sato, T. Shibata, S. Komiya, M. Hayashi, Y. Matsueda, H. Imamura, T. Hayase, H. Morioka, H. Tabira, J. Yamamoto and Y. Kowata, *Tetrahedron: Asymmetry*, 2003, **14**, 185–188.
5. (a) E. Havinga, *Biochim. Biophys. Acta*, 1954, **13**, 171–174; (b) D. K. Kondepudi and K. Asakura, *Acc. Chem. Res.*, 2001, **34**, 946–954.
6. D. K. Kondepudi, K. L. Bullock, J. A. Digits, J. K. Hall and J. M. Miller, *J. Am. Chem. Soc.*, 1993, **115**, 10211–10216.
7. D. J. Durand, D. K. Kondepudi, P. F. Morelra Jr. and F. H. Quina, *Chirality*, 2002, **14**, 284–287.
8. (a) D. K. Kondepudi, J. Laudalio and K. Asakura, *J. Am. Chem. Soc.*, 1999, **121**, 1448–1451; (b) K. Asakura, T. Soga, T. Uchida, S. Osanai and D. K. Kondepudi, *Chirality*, 2002, **14**, 85–89; (c) K. Asakura, Y. Nagasaka, M. Hidaka, M. Hayashi, S. Osanai and D. K. Kondepudi, *Chirality*, 2004, **16**, 131–136.
9. M. Sakamoto, N. Utsumi, M. Ando, M. Saeki, T. Mino, T. Fujita, A. Katoh, T. Nishio and C. Kashima, *Angew. Chem., Int. Ed.*, 2003, **42**, 4360–4363.
10. J. H. E. Cartwright, J. M. Garcia-Ruiz, O. Piro, C. I. Sainz-Diaz and I. Tuval, *Phys. Rev. Lett.*, 2004, **93**, 035502/4.
11. C. Viedma, *J. Crystal Growth*, 2004, **261**, 118–121.
12. C. Viedma, *Phys. Rev. Lett.*, 2005, **94**, 065504/4.
13. R. Plasson, D. K. Kondepudi, H. Bersini, A. Commeyras and K. Asakura, *Chirality*, 2007, **19**, 589–600.
14. C. Viedma, *Astrobiology*, 2007, **7**, 312–319.
15. J. H. E. Cartwright, O. Piro and I. Tuval, *Phys. Rev. Lett.*, 2007, **98**, 165501/4.
16. (a) W. Ostwald, *Z. Phys. Chem.*, 1897, **22**, 289–302; (b) R. Boistelle and J. P. Astier, *J. Cryst. Growth*, 1988, **90**, 14–30.
17. W. L. Noorduin, T. Izumi, A. Millemaggi, M. Leeman, H. Meekes, W. J. P. van Enckevort, R. M. Kellogg, B. Kaptein, E. Vlieg and D. G. Blackmond, *J. Am. Chem. Soc.*, 2008, **130**, 1158–1159.

CHAPTER 7
Outside Earth: Meteorites and Comets

7.1 Introduction

Either chance mechanisms, spontaneous symmetry breaking, or any other kind of local asymmetry, has a 50% probability of occurring and will give different handedness in different locations. On the other hand, the weak force is a universal chiral influence—it will produce the same chiral imprint throughout the universe.

Looking beyond Earth might provide essential clues about the distinction between, on the one hand, the so-called chance mechanisms (Section 2.2, *Chance Theories*) and local deterministic (or regional) mechanisms, and on the other hand, the universal deterministic mechanisms (Section 2.3, *Deterministic Theories*). The only weakness in this assumption is that it is necessary to define how large the region is in which a local deterministic mechanisms operates. If the region is very large, for example the size of the solar system, observation of the imprint of chirality in other regions may become a very difficult task, or at present simply impossible. For the time being, examining pieces of evidences coming from outside the Earth is an easier target, and this also can provide important clues as to how homochirality might have developed.

7.2 Meteorites

The Murchison meteorite is one of the most famous and most studied sources of evidence arriving from outer space. It is a CM2 carbonaceous chondrite (CM named after the Mighei meteorite in Ukraine), a rare and ancient type of stony meteorite which, in addition to silicates, oxides and sulfides, *etc.*, most distinctively also contains water. In addition it carries significant amounts of carbon (2%), consisting mostly of macromolecular carbon (70–80%), carbonate minerals (2–10%), and 10–20% of low molecular weight organic compounds,

---

The Origin of Chirality in the Molecules of Life
Albert Guijarro and Miguel Yus
© Albert Guijarro and Miguel Yus, 2009
Published by the Royal Society of Chemistry, www.rsc.org

Table 7.1 Soluble organic compounds in the Murchison meteorite.[a]

Class	Concentration (ppm)[b]	Compounds identified[c]
Aliphatic hydrocarbons	>35[c], >10[d]	140
Aromatic hydrocarbons	15–28[c], >10[d]	87
Dicarboxylic acids	>30[c], >10[d]	17
Carboxylic acids	>300[c], >100[d]	20
Pyridine carboxylic acids	>7[c]	7
Dicarboxyimides	>50[c]	3
Sulfonic acids	67[c], >100[d]	4
Amino acids	60[c], >10[d]	74
Amines	8[c], >1[d]	10
Amides	>10[c], >10[d]	4
Hydroxy acids	15[c], >10[d]	7
Aldehydes, ketones	>10[d]	–
Phosphonic acids	>1[d]	–
N-heterocycles	>1[d]	–
Purines, pyrimidines	>1[d]	7
Polyols, sugar-related compounds	>250 nmol/g[e]	>30

[a] Analyses from different fragments of the Murchison Meteorite.
[b] Extraction with aqueous HCl at 100 °C.
[c] Reference 8.
[d] Reference 9.
[e] Reference 5.

known as the soluble fraction (Table 7.1). This fraction contains a wide variety of amino acids, including α-amino acids such as glycine, alanine, leucine, proline, glutamic acid, aspartic acid, valine and serine, and a range of non-proteinogenic α-, β-, γ- and δ-amino acids.[1] Also present are hydroxy acids,[2] carboxylic, sulfonic and phosphonic acids,[3] amines, aliphatic and aromatic hydrocarbons and fullerenes, as well as other significant organic compounds, including nitrogenated heterocycles (*e.g.*, purines, and pyrimidines such as xanthine, adenine and uracil),[4] and sugars or sugar derivatives (C_2 to C_6 polyols, *e.g.*, ribitol, glucitol and its isomers, ribonic, gluconic acid and its isomers, *etc.*),[5] the sugars and sugar derivatives devoid of chiral studies as far as we know.

The Murchison meteorite exploded into fragments over the town of Murchison, Australia, on September 1969. About 82 kg of fragments were recovered. Subsequent analysis by NASA scientists revealed the presence of a number of amino acids commonly found in proteins, and others which were not present in terrestrial life. Initial studies did not point to a chiral bias in the distribution of amino acids in the Murchison meteorite. The meteorite also contained hydrocarbons, which appeared abiogenic in character due to their enriched heavy carbon isotopic composition, confirming the extraterrestrial origin of the organic compounds.[6] In general, terrestrial organic matter tends to be depleted in the stable heavier isotopes, such as ^{13}C, ^{15}N and ^2H, in comparison with interstellar isotopic proportions due to biosynthetic isotope fractionation.[7]

More than 76 amino acids have been identified in all in the Murchison meteorite. Since its arrival there have been hints of a slight enantiomeric excess of the L-series of amino acids, but the scientific community has remained skeptical about these findings; near-racemic samples have strongly pointed to the possibility of biological contamination.[10] However, in 1997 further analysis reported an excess of 7–9% in the left-handed versions of four amino acids,[11] a result later confirmed by an independent group.[1] In this work a different strategy was adopted to minimize contamination issues and epimerization artifacts at the C-α. The study focused on nonterrestrial α-branched amino acids, which had the advantage of being nonracemizable and not easily contaminated biologically. In particular, the four stereoisomers of 2-amino-2,3-dimethylpentanoic acid, *i.e.*, DL-α-methylisoleucine and DL-α-methylalloisoleucine, none of which are known in the biosphere, showed that the L-enantiomer occurred in 7.0% and 9.1% excess, respectively. Similar results were obtained for two further α-methylamino acids, DL-isovaline (rare on Earth) and DL-α-methylnorvaline (unknown on Earth) (Figure 7.1).[11]

In general, the common amino acids containing asymmetric centers, even those epimerizable at C-α are not racemic and they all show an excess of the L-enantiomer in varying degrees,[1] the magnitude of which has not always been confirmed by other groups.[12] Among the racemic or near-racemic amino acids, DL-α-amino-*n*-butyric acid (common on Earth) and DL-α-amino-*n*-valeric acid (rare on Earth) were present, both carrying an easily epimerizable α-hydrogen.

Isotopic analysis of ^{15}N in individual amino acids found in the Murchison meteorite are more reliable than those previously reported for ^{13}C depletion, due to higher isotopic enrichment. These analyses showed moderate to extreme

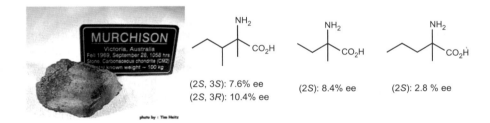

Figure 7.1 A fragment of the Murchison meteorite. Several nonracemic (L)-α-branched amino acids were found in this meteorite, the reported *ee* being obtained by chiral GC–MS after derivatization. Virtually unknown in the biosphere, these amino acids are not readily racemizable, nor are they easily affected by biological contamination or processing, and are good candidates for analysis of enantiomeric excess. The four stereoisomers of 2-amino-2, 3-dimethylpentanoic acid, *i.e.*, DL-α-methylisoleucine (2*S*, 3*S*) and DL-α-methylalloisoleucine (2*S*, 3*R*), all of which are nonterrestrial, showed enantiomeric enrichment of the L-enantiomer (7.0 and 9.1 % *ee* on average, respectively; the values in Figure 7.2 are the highest single experimental figures obtained). Similar results were obtained for two other α-methyl amino acids, DL-isovaline (rare on Earth) and DL-α-methylnorvaline (unknown on Earth). The results have been corroborated by independent groups and from other meteorites (*e.g.*, Murray), and this can be considered to be hard evidence of the existence of enantiomeric excess in meteorites. (Picture from http://www.meteorman.org, with the author's permission.)

enrichment in ^{15}N relative to their terrestrial counterparts in both enantiomers, confirming the extraterrestrial source of these amino acids.[1]

Similar conclusions can be drawn using a different mode of reasoning. In general the families of organic compounds found in meteorites follow a characteristic abiotic synthetic pattern: a general decrease in abundance with increasing carbon number within one class of compounds, and many or all of the possible isomers being present at a given carbon number (*e.g.*, within the sugar series, or the amino acids). This fractionation and lack of selection are indicative of abiotic synthesis from simple precursor molecules.

The Murchison meteorite is not unique, however. L-Enantiomeric enrichment is also present in other carbonaceous chondrites.[12,13] Almost identical amino acid compositions have also been found in other CM2 meteorites, including Murray, which fell near Murray, Kentucky, USA, in 1950, and more recently in the Antarctic meteorite LEW90500 found resting in ice after tens of thousands of years, but still very low in contamination with terrestrial amino acids.[14] The total abundance of amino acids in these meteorites is significantly lower, with ~7760 parts per billion (ppb) in LEW90500 and 11 600 ppb in Murray, compared with ~15 300 ppb in Murchison, based on bulk samples.

Regarding other types of meteorites, CI1 carbonaceous chondrites (CI named after the Ivuna meteorite) such as Orgueil, which fell near the French town of Orgueil in 1864 and was studied by Pasteur himself, and the Ivuna meteorite itself, which landed near Ivuna, Tanzania, in 1938, have a different and simpler amino acid composition compared to CM2.[14] While CM2 carbonaceous chondrites contain a wide variety of complex amino acids, the CI type have a simpler composition, predominantly glycine and β-alanine, each present in significant quantities but unfortunately devoid of stereogenic centers. Nevertheless, the trace amounts of aspartic and glutamic acids present in the Orgueil and Ivuna metorites appear to be enriched in the L-enantiomer.

Assessment of the way these compounds were formed provides important information about the chemical reactions which took place in the early solar system. Amino acids from Murchison, Murray and LEW90 500 are thought to have been formed by the Strecker reaction, a synthetic pathway involving precursor molecules such as hydrogen cyanide, ammonia, aldehydes and ketones, which from their isotopic make-up are probably of interstellar origin. These molecules would have been trapped on the meteorite's parent body, presumably an asteroid,[15] and involved in reactions under aqueous conditions. Recent research indicates that certain dark asteroids within the main asteroid belt could be the immediate source of CM meteorites. In contrast, the Orgueil and Ivuna meteorites contain a strikingly simple amino acid mixture, primarily glycine and β-alanine, with traces of only a few other aminoacids. The amino acids in these meteorites are likely to have formed from a limited set of precursor components such as hydrogen cyanide, ammonia and cyanoacetylene, all of which have been detected in comet tails. This implies a different type of reactions and parentage than Murchison and Murray, possibly an extinct comet.[14]

Conjugate addition of ammonia to α,β-unsaturated nitriles followed by hydrolysis has been suggested as a source of the β-amino acids abundant in CI meteorites. Figure 7.2 shows the analyses performed on the aqueous extracts of these renowned meteorites, including Murchison. After chiral derivatization with *o*-phthalodialdehyde/*N*-acetyl-L-cysteine, and serpentine and blank correction, HPLC analyses indicate the presence of 11 identifiable amino acids, some of them as resolved enantiomers (aspartic acid, glutamic acid and alanine). Aspartic and glutamic acid L-enrichment was detected in Orgueil, Ivuna,

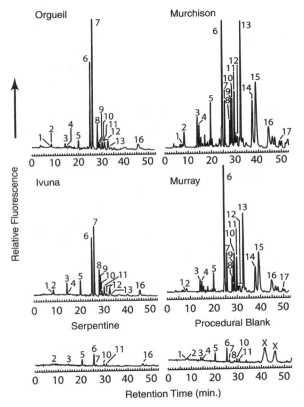

Figure 7.2 Meteorite amino acids. HPLC chromatograms of the *o*-phthalodialdehyde/ *N*-acetyl-L-cysteine derivatization of amino acids in the aqueous extracts from the CI carbonaceous chondrites Orgueil and Ivuna, the CM chondrites Murchison and Murray, and the serpentine (silicate constituent of stony meteorites) and procedural blanks. Peaks were identified as follows: 1: D-aspartic acid; 2: L-aspartic acid; 3: L-glutamic acid; 4: D-glutamic acid; 5: DL-serine; 6: glycine; 7: β-alanine; 8: γ-ABA; 9: D,L-β-AIB; 10: D-alanine; 11: L-alanine; 12: D,L-β-ABA; 13: AIB; 14: D,L-α-ABA; 15: D,L-isovaline; 16: L-valine; 17: D-valine; and X: unknown. (ABA: aminobutiric acid. AIB: aminoisobutiric acid). Resolved amino acids (aspartic and glutamic acids, alanine) were found to be enriched in the L-form (6 cases) or near racemic (6 cases) in the four meteorites. (From Reference 14, with authorization).

Murray and Murchison, with the exception of racemic alanine in Orgueil and near racemic L-alanine in Murray. [Serpentine, approx. $(Mg,Fe)_3Si_2O_5(OH)_4$, is a phylosilicate and a mineral constituent of the stony meteorites used in blank correction protocols].

Carbonaceous chondrites are the most primitive and unaltered type of meteorites known. They date from around 4500 million years ago, about as old as the solar system itself. This is long before the appearance of life on Earth, which is believed to have arisen some 3800 million years ago. Some of the amino acids from CM meteorites have strange isotopic signatures, which tends to suggest that their component elements did not originate within our solar system. These amino acids are believed to represent actual interstellar matter that was trapped in the meteorite more than 4500 million years ago.[16] Despite the fact that the constituents might be very different in origin, for example from asteroids or the remains of a comet, the amino acids, seldom racemic, are enriched in the L-form (Figures 7.2 and 7.3).[14,17] The amount by which different fragments differ in their enantiomeric excess is only of relative importance—it depends on the processing history of each stone and parent body. The crucial question, from which our attention should not be diverged, is why there is any enantiomeric excess at all. Any deviation from the racemic condition is a very significant finding and one which demands a satisfactory explanation. There is no question that meteorites, in particular carbonaceous chondrites, provide a crucial piece of information which must be accommodated within any theory of the origins of biomolecular homochirality. They represent the only direct experimental evidence of chemical evolution in the early solar system and provide a natural record of its organic prebiotic chemistry.[18] From this extraterrestrial evidence, prebiotic nonracemic conditions seem to have been present at the actual time when life evolved.

7.3 Comets

Comets are another kind of small celestial body able to render important information regarding molecular chirality from remote parts of the solar system. A comet's nucleus is a lumpy conglomerate in which dust is held together by a large amount of ice. Comets are thought to fall gravitationally from the Kuiper belt or the Oort cloud (large reservoirs of comets near or beyond the boundaries of the solar system) in the direction of the sun, induced by gravitational perturbations. If they are captured by the sun, they orbit with great eccentricity in their fall and decay over a relatively short lapse of time (between a few centuries and a few millennia), their components being vaporized close to the sun and producing enormous tails of volatiles (Figure 7.4, *top*). They are thought to preserve the most pristine and unaltered material in the solar system, dating back to the pre-solar or solar nebulae, the huge cloud of gas and dust out of which the solar system was formed.[20]

Cometary dust is rich in organics. The spectra of comet tails reveal the presence of alcohols, aldehydes, ketones, acids, amino acids, and *N*-heterocyclic

Outside Earth: Meteorites and Comets

Figure 7.3 Some meteorites are among the oldest sources of material evidence we have, dating back more than 4500 million years, long before life began (*ca.* 3800 million years ago) and nearly as old as the sun itself. CM2 meteorites such as Murchison or Murray are believed to be of asteroidal origin, whereas the Ivuna, a CI1 type carrying a different amino acid composition, had an unrelated origin, probably an extinct comet. Despite the different genesis and location of their parent bodies in the solar system, such as the asteroid belt or comets, the α-amino acids present are found to be slightly enantiomerically enriched in the L-enantiomer. (Pictures of fragments of meteorites from http://fernlea.tripod.com (Murray) and http://www.asahi-net.or.jp/~ug7s-ktu (Ivuna) are presented with permission, and small bodies of the solar system are from Reference 19 with authorization.)

compounds, although the exact composition of these molecules has not been explicitly determined.[21] At present, comets are the subject of intense research, providing a means of looking back in time to the origins of the solar system. Indeed, cometary material represents the closest we can get to the conditions existing when the sun and our solar system were born. With this in mind the ESA (European Space Agency) is undertaking one of the most challenging space missions so far attempted, the Rosetta mission.[22] In March 2004 the cometary

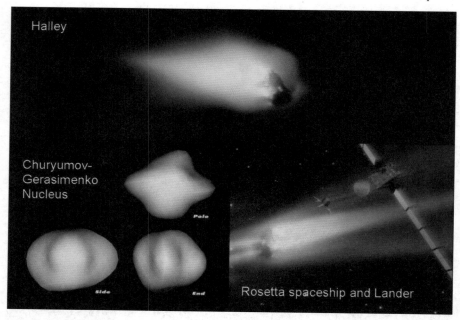

Figure 7.4 The Halley comet showing its nucleus as seen from the ESA Giotto spacecraft. In 1986 this passed close to the nucleus of Halley, providing the first evidence of organic material in a comet. In a very ambitious ESA project, culminating in an actual landing on a comet, the Rosetta spaceship began its 10-year expedition in March 2004 towards the outer reaches of the solar system. Rosetta will reach its main target, Comet 67P/Churyumov–Gerasimenko in May 2014 (*lower left*: 67P/C-G nucleus of *ca.* 5 km × 3 km as seen by the Hubble space telescope). Following this, a robotic lander will detach from the Rosetta orbiter and descend on to the surface of the comet, performing chiral GC–MS analyses *in situ* (*lower right*, an ESA artist's impression). A variety of organic molecules, such as alcohols, ketones and amino acids, are anticipated. The sign and magnitude, if any, of the enantiomeric excess of some fractions (*e.g.*, those containing α-amino acids) is awaited with great anticipation. (Pictures with approval from the European Space Agency multimedia gallery: http://www.esa.int/esaCP/index.html.)

probe Rosetta was launched in order for the first time to make physical contact with a comet, following a 10-year journey. By the end of the summer of 2014, the robotical lander Philae will detach from the orbiter of the Rosetta spacecraft and land on the surface of comet 67P/Churyumov–Gerasimenko in an attempt to separate and identify cometary organic compounds *in situ* using GC–MS. In an experiment called COSAC (Cometary Sampling and Composition), chiral organics will be separated into their enantiomers by means of capillary columns coated with different types of chiral and nonchiral stationary phases, followed by MS analysis using a protocol which minimizes coelution errors and other artifacts. Nonvolatile compounds such as amino acids will be derivatized first and analyzed in a similar manner. In a separate experiment another evolved gas analyzer, Modulus Ptolemy, will obtain accurate measurements of the isotopic ratios of lighter elements. Fundamental aspects of prebiotic chemistry, such as the water in our oceans, may well contain the clues of its origin written in isotopic code. As it approaches the sun Rosetta will orbit the comet over the course of an entire year, recording changes in its activity. As the comet's ice evaporates, instruments on board the orbiter will study the dust and gas particles which surround it, forming streaming tails loaded with organics. Little is known about this comet, despite it being a large dirty snowball that regularly visits the inner solar system, orbiting the sun once every 6.6 years (Figure 7.4, *lower left* and *right*). The goals of the mission are far reaching, and scientists of many disciplines await its outcome with eager anticipation. The data from the state of the art instruments will be crucial in testing the different hypotheses, not only regarding the origins of biomolecular homochirality, including early seeding of Earth with chiral organic molecules,[23] but also the origins of life itself.

References

1. M. H. Engel and S. A. Macko, *Nature*, 1997, **389**, 265–268.
2. J. R. Cronin and S. Pizzarello, *Geochim. Cosmochim. Acta*, 1986, **50**, 2419–2427.
3. G. W. Cooper, W. M. Onwo and J. R. Cronin, *Geochim. Cosmochim. Acta*, 1992, **56**, 4109–4015.
4. P. G. Stoks and A. W. Schwartz, *Geochim. Cosmochim. Acta*, 1982, **46**, 309–315.
5. G. Cooper, N. Kimmich, W. Belisle, J. Sarinana, K. Brabham and L. Garrel, *Nature*, 2001, **414**, 879–883.
6. (a) I. Gilmour and C. Pillinger, *Meteorites*, 1997, **27**, 224–225; (b) G. Yuen, N. Blair, D. J. Des Marais and S. Chang, *Nature*, 1984, **307**, 252–254; (c) M. H. Engel, S. A. Macko and J. A. Silfer, *Nature*, 1990, **348**, 47–49.
7. M. L. Fogel and L. A. Cifuentes, in *Organic Geochemistry, Principles and Applications*, M. H. Engel and S. A. Macko ed., Plenum, New York, 1993, 73–98.
8. (a) S. Pizzarello, Y. Huang, L. Becker, R. J. Poreda, R. A. Nieman, G. Cooper and M. Williams, *Science*, 2001, **293**, 2236–2239; (b) J. R. Cronin, S. Pizzarello and D. P. Cruikshank, in *Meteorites and the Early Solar System*, ed. J. F. Kerridge and M. S. Matthews, University of Arizona Press, Tucson, 1988, 819–857.
9. (a) J. Podlech, *Cell. Mol. Life Sci.*, 2001, **58**, 44–60; (b) J. R. Cronin, Clues from the Origin of the Solar System: Meteorites, in *The Molecular Origins of Life: Assembling Pieces of the Puzzle*, ed. A. Brack, Cambridge University Press, Cambridge, UK, 1998, 119–146.
10. (a) M. H. Engel and B. Nagy, *Nature*, 1982, **296**, 837–840; (b) M. H. Engel and B. Nagy, *Nature*, 1983, **301**, 496–497.
11. J. R. Cronin and S. Pizzarello, *Science*, 1997, **275**, 951–955.
12. J. R. Cronin and S. Pizzarello, *Adv. Space Res.*, 1999, **23**, 293–299.

13. S. Pizzarello, M. Zolensky and K. A. Turk, *Geochim. Cosmochim. Acta*, 2003, **67**, 1589–1595.
14. P. Ehrenfreund, D. P. Glavin, O. Botta, G. Cooper and J. L. Bada, *Proc. Natl. Acad. Sci. USA*, 2001, **98**, 2138–2141.
15. T. Hiroi, C. M. Pieters, M. E. Zolensky and M. E. Lipschutz, *Science*, 1993, **261**, 1016–1018.
16. F. L. Plows, J. E. Elsila, R. N. Zare and P. R. Buseck, *Geochim. Cosmochim. Acta*, 2003, **67**, 1429–1436.
17. M. H. Engel and S. A. Macko, *Precambrian Res.*, 2001, **106**, 35–45.
18. S. Pizzarello, *Chem. Biodivers.*, 2007, **4**, 680–693.
19. D. Yeomans, *Nature*, 2000, **404**, 829–832.
20. J. M. Greenberg, *Sci. Am.*, 1984, **250**, 124–135.
21. M. N. Fomenkova, *Space Sci. Rev.*, 1999, **90**, 109–114.
22. http://www.esa.int/export/esaMI/Rosetta.
23. U. Meierhenrich, H. -P. Thiemann and H. Rosenbauer, *Chirality*, 1999, **11**, 575–582.

CHAPTER 8
Other Local Deterministic Theories

8.1 Introduction

A number of local deterministic theories have already been described in other chapters related, for example, to the asymmetry of the Earth's hemispheres in their interaction with solar light (Section 4.3.1, *Circularly Polarized Light: Circular Dichroism*), due to the direction of the magnetic field (Section 4.3.2, *The Magnetochiral Effect*), or the Coriolis effect (Section 4.4, *Fluid Dynamics: Vortex Motion*).

8.2 Chiral Crystals and Faces on Crystals

The possible actions of enantiomorphic inorganic crystals or crystal faces of the Earth's minerals and rocks as chiral inducers for organic compounds has long ago been suggested for a number of minerals, including quartz,[1] calcite,[2] gypsum[3] and others.[4] Diastereomeric interactions can take place between a chiral molecule and a chiral crystalline surface. As a consequence of this, stereoselective adsorption may develop, allowing a certain level of enantiomeric discrimination in a racemic sample. On the other hand, the presence of a chiral surface can also activate prochiral compounds in an asymmetric (and presumably catalytic) manner, resulting in enantioselective synthesis. Chiral transmission is therefore possible in any of these situations, provided that a chiral array of molecules on some exposed surface is available.

8.2.1 Quartz

α-Quartz is by far the most common non-centrosymmetric (chiral) mineral in Nature, and it has deservedly attracted much interest regarding its possible role as a chiral inductor in prebiotic chemistry. It is also the most stable form of SiO_2 at room temperature and atmospheric pressure, until the temperature reaches 573 °C, at which there is a phase transition to β-quartz. In addition to α- and β-quartz, there are many other varieties of forms of SiO_2, many of them showing only small differences in energy between the phases.[5] Mineralogical data tell us that α-quartz belongs to the trigonal–trapezohedral crystal class, and to $P\,3_121$ (left-handed crystals) or $P\,3_221$ (right-handed crystals) space group symmetry. These are chiral crystal structures, and quartz is therefore an optically active material.[6,7] The idealized crystal of α-quartz, in its normal (or prismatic) habit, has a single axis of three-fold symmetry along the prism (trigonal or optical axis) and three perpendicular axes of two-fold symmetry (diagonal axes) spaced 120° apart, which are polar axes (D_3 point group). These diagonal axes are described as polar, since a definite sense can be assigned to them (different faces are shown at each end of the polar axis). Since this is the case, it follows that the crystal exhibits hemimorphism. When a beam of plane-polarized light is transmitted along the trigonal axis a rotation of the plane of polarization occurs, the degree of rotation depending on the thickness or the distance traversed in the material. The sense of rotation can be used to differentiate between the two naturally occurring forms of α-quartz: the levorotatory, (−)- or *l*-quartz, and the dextrorotatory, (+)- or *d*-quartz. Fortunately, since α-quartz displays hemimorphism it is often possible to recognize the chiral sense of the enantiomorphous crystal simply by examination of the outward shape of the crystal. This brought about the alternative crystallographic terminology of *left*-quartz or *l*-quartz, and *right*-quartz or *r*-quartz, which has become widely accepted. Nevertheless, if we adopt the convention that the direction of rotation is that seen by an observer looking back toward the source of light—which is the commonly accepted sense used in polarimetry—a given crystal is either *right* or *left* from both the crystallographic and the optical standpoint.

The α-quartz crystal is a hexagonal prism with six cap faces at each end. Frequently only one end of the crystal is visible, the other end being embedded in the rock matrix. Following the original Pasteur drawings (Figure 8.1), the prism faces are designated as *m* faces and the cap faces as *r* and *z* faces. The *r* faces are also called major (or positive) rhombohedral faces and the *z* faces are minor (or negative) rhombohedral faces. Bordered by the prism faces and the rhombohedral faces, and depending on the habit, there are the hemihedral faces, represented as *x* and *s*. The presence of polar axes and hemihedral faces implies the absence of a centre of symmetry, and therefore the existence of enantiomorphism. Left- and right-handed crystals can be distinguished by the position of the hemihedral faces (Figure 8.1). This can be achieved by holding the crystal vertically and rotating it around the trigonal axis until the major rhombohedron (*r*) faces the observer. If the small faces of the trigonal trapezohedron (*x*) and the trigonal bipyramid (*s*) are at the left lower corner of the major rhombohedron (*r*), then by convention the crystal is a *left*-handed crystal and is levorotatory (Figure 8.1, *left*). If these two faces are on the lower right of the major rhombohedron (*r*), then the crystal is *right*-handed and is dextrorotatory (Figure 8.1, *right*).

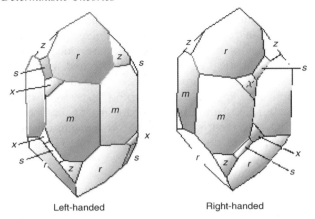

Figure 8.1 The mineral quartz occurs either as *left*- or *right*-handed specimens, reflecting the chirality of its internal helical structure. In hemihedral crystals, such as those in the picture (drawn originally by Louis Pasteur), the appearance of hemihedral faces (x, s) allows the assignment of chirality by simple inspection of their outward shape.

In spite of their well defined external morphology and habit, many natural samples of quartz crystals are internally "twinned", in other words they exhibit both left- and right-handed domains (twinning refers to the existence of various homogeneous domains orientated in different directions within a crystal).[8] In a slice of a twinned crystal the left- and right-handed portions can still be distinguished using polarized light. On the other hand cultured quartz can be grown artificially, free from twinning and therefore internally homochiral, a condition required for most technical uses in electronics and other scientific applications.

The example of quartz should not however be taken as a general method of identifying chiral molecular structures from the shape of the crystal, but rather as a less usual manifestation which happens in the case of a very common mineral. The occurrence of hemihedral faces, or hemimorphism, in quartz and other hemihedral crystals such as sodium ammonium tartrate (Chapter 1, *Introduction and Historical Background*) allows direct assignment of chirality, since these motifs themselves exist in a discernible chiral environment; those oriented in one sense correspond to one enantiomorphous crystal, while those oriented in the opposite sense identify the other enantiomorphous crystal. Unfortunately such assignment of handedness is only possible in crystals displaying hemihedral faces. In the case of holohedral crystals, which are those having the greatest symmetry within their crystal class, each face is accompanied by a parallel face on the opposite side of the crystal, and direct assignment of external chirality is not possible.

The discovery of asymmetric adsorption and asymmetric catalysis involving optically active quartz crystals, along with the mistaken belief that there was a preponderance in Nature of *l*-quartz, led some authors to deduce that this source of asymmetry could have played an important role as a source of chiral bias in Nature (Section 5.4, *The Yamagata Cumulative Mechanism*).[9] Later, data from a more complete range of samples revealed that *d*-quartz and *l*-quartz were within statistical error equally distributed worldwide.[10]

Regarding the internal structure of quartz,[11] the thermodynamically most stable phase at ambient temperature (α-quartz) consists of a three-dimensional network of tetrahedral [SiO$_4$] units, forming interlinked helical chains (Figures 8.2 and 8.3). Each [SiO$_4$] tetrahedron belongs to two nonequivalent helical chains, but running in parallel and displaying similar helicity. The helixes in any one

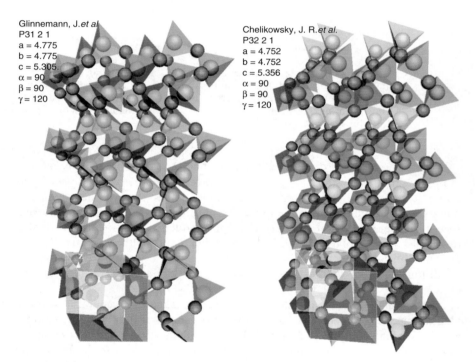

Figure 8.2 Among the many polymorphs of silica, α-quartz (shown) is the most stable crystalline phase at room temperature (Si = yellow; O = red). This covalent crystal is composed of a network of slightly deformed [SiO$_4$] tetrahedral building blocks, with two slightly nonequivalent Si–O bond lengths in each tetrahedron. This asymmetry is a consequence of the helical twisting of the network of chains, endowing chirality to the entire crystal structure and in turn producing enantiomorphic crystals. On the *left*, the structure of the *l*-α-quartz, displaying a helical arrangement of chains along the trigonal axis (z), is from Reference 7. On the *right*, the enantiomorphic structure of *r*-α-quartz, viewed from a similar perspective and displaying the opposite helicity in the chains, is from Reference 6. These crystals display the opposite optical rotation when a beam of plane-polarized light is projected along the trigonal (or optical) axis. The spatial groups, in addition to the parameters of the unit cell, are shown. CIF files of quartz can be obtained from the free Crystallography Open Database: http://www.crystallography.net.

crystal can be either *right*-handed or *left*-handed, so that individual crystals possess non-superimposable mirror image structures which are manifested in their outward shape. As mentioned earlier, the crystals can often be easily separated by hand, provided the specimens are not twinned and that they display well developed hemimorphism.

8.2.2 Calcite, Gypsum, Clay Minerals and Others

Even when they are within an achiral body, centrosymmetric crystals can exhibit a rich variety of chiral surfaces. As a result of the inversion centre of the crystal, each of these chiral surfaces has a corresponding surface of opposite chirality on the opposing side of the crystal, so the net chiral effect is in balance. However, taken individually these surfaces are examples of local chirality, and many rock-forming minerals display crystal growth faces with these characteristics. Calcite ($CaCO_3$) is one such example. Unlike quartz, calcite crystals (trigonal system, hexagonal scalenohedral class, space group $R-3c$) in the common scalenohedral-shaped crystals, or "dogtooth" habit, are centrosymmetric, displaying no chirality as a whole but forming mirror image crystal surfaces (Figure 8.3). These enantiomorphic surfaces exhibit selective adsorption of some chiral molecules, such as amino acids, *e.g.*, aspartic acid. Faces of similar handedness selectively adsorb D-aspartic acid, while the L-enantiomer selectively adheres to mirror-related faces on the same crystal, with *ee* of the order of 0.5%.[2]

Monoclinic gypsum ($CaSO_4.2H_2O$) is another centrosymmetric crystal (monoclinic–prismatic class, space group $A\ 2/a$) which behaves similarly. The crystal growth morphology of gypsum is strongly influenced by the presence of chiral molecules which selectively adsorb on specific chiral faces. The growth rate on these individual faces is affected, resulting in preferential deposition on other faces and displaying macroscopic asymmetric crystal habit.[3] Feldspars, which make up the greatest percentage of the minerals present in the Earth's crust, also provide promising templates for the selective adsorption and organization of chiral species on their chiral surfaces.[12] But perhaps the most striking candidates for enantioselective processing during prebiotic evolution are the clays. The possible implication of clay minerals in prebiotic transformations has been postulated and studied over many years,[13] and has shown that some clay minerals which caused swelling were able to perform stereoselective catalysis, evoking a pseudo-enzymatic role.[14] However, the issue which has caused most concern is asymmetric stereoselectivity, allegedly displayed by some clay minerals in certain biochemically-related processes such as the polymerization and adsorption of amino acids.[15] These results failed to demonstrate adequate reproducibility and were quickly refuted,[16] but further reports of enantioselective physical or chemical processes on clay surfaces still appear sporadically in the literature.[17,18]

The clay minerals are an important component within the more general phyllosilicate group, which are characterized by a layered structure.[19] In a

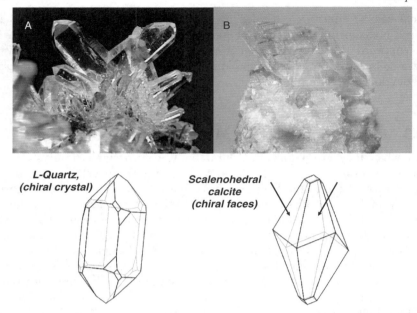

Figure 8.3 (A) Quartz (SiO$_2$) is the second most abundant mineral in the Earth's crust, the α-form being the stable phase at ambient temperature. It is also the most common non-centrosymmetric (chiral) mineral on Earth, enantiomorphic crystals being equally populated and randomly distributed. In illustration (A) two large quartz crystals have grown orthogonal to each other in a V-shape, forming "Japan-law twins". In this crystalline association both crystals share a mirror plane and are enantiomorphic ("Japan–Brazil-law twins"). The minor crystal facets in each α-quartz subunit follow either a left- or a right-handed screw sequence, better seen in the model of left-handed α-quartz below, showing opposite chiral structures in crystals grown side by side from the same source. (B) On the other hand, calcite (CaCO$_3$), the third most common mineral in the Earth's crust, forms achiral crystals with mirror-image crystal faces disposed in pairs (indicated with arrows in scalenohedric calcite). These chiral faces show significant enantioselection in the adsorption of chiral amino acids. (Sources: Quartz (Japan-law twins), http://mineral.galleries.com; dogtooth scalenohedric calcite crystal on calcite matrix, courtesy of C. Tucker: http://www.christuckerminerals.com; drawings from http://webmineral.com.)

simplified description of the structure, [SiO$_4$] tetrahedra are linked by three shared oxygens in a two-dimensional plane which produce a sheet-like structure of empirical formula (Si$_2$O$_5^{-2}$)$_n$. All the remaining unshared vertices carrying the anionic oxides are pointing in the same direction and on the same side of the sheet. This is the silicate layer. Typically, these sheets are then connected to layers of aluminum hydroxide, which has a basic structure comprising sheets of linked octahedrons of aluminum hydroxide (as in Gibbsite), of the general empirical formula (Al$_2$(OH)$_4^{+2}$)$_n$. In 1:1 clays the unshared vertex from the tetrahedral sheet is linked to one side of the octahedral sheet. Alternatively, two layers of silicate may be linked to the two sides of the octahedral sheet, forming 2:1 clays. Additionally, isomorphic substitution of Si(IV) by tetrahedral Al(III), and of Al(III) by octahedral Mg(II), allows the formation of anionic sheets which are counterbalanced by a variety of other cations occupying interlayer positions. These layers are weakly bonded between them and often contain large amounts of water, and can host other neutral molecules trapped between the sheets.

Clays exhibit high adsorption properties, ion exchange and catalytic activity, including peptide formation,[20] and are normally formed by weathering or alteration of other silicate minerals. Kaolinite, essentially $Al_2Si_2O_5(OH)_4$, (triclinic–pedial class, space group $P\,1$) is a non-centrosymmetric structure of the 1:1 clay type, and is a common white clay mineral formed by weathering of feldspars. In kaolinite the interlayer is sandwiched between silicate sheets $[Si_2O_5]$ on one side and aluminum oxide/hydroxide layers $[Al_2(OH)_4]$ on the other side, producing specific guest orientations,[21] with two enantiomorphic arrangements of the interlayer space corresponding to the two enantiomorphic structures.[22] These are named A and B forms, in relation to the (a, b, c) axes of the unit cell or trihedron (three vectors with a common vertex), which may be either direct or indirect (forms A and B, respectively). Quantum calculations show that the adsorption of an ethyliminium cation ($CH_3CH=NH_2^+$) on a crystal of kaolinite of a given handedness, e.g., form A, occurs in a manner which implies the addition of CN^- preferentially from one side, producing in the example indicated L-aminopropionitrile, a precursor of L-alanine by hydrolysis: $CH_3CH=NH_2^+ + CN^- \rightarrow$ L-$CH_3CH(NH_2)CN$. The formation of the L-alanine precursor is favored over that of the D-alanine precursor by $0.36\,kcal\,mol^{-1}$.[23] The same conclusion is obtained for the adsorption of alanine itself on kaolinite, but with lower differentiation between the binding energies of the L- and D-forms, either as the positive ion or the zwitterion. Similarly to the reported studies on quartz, weak force-implemented quantum calculations revealed a vanishingly small parity-violating energy difference (ΔE_{pv}) favoring the A-form of kaolinite,[24] which can now be dismissed as obsolete, or in any event inconclusive.

Montmorillonite [essentially $Al_2(Si_2O_5)_2(OH)_2.nH_2O$] is another common clay mineral of the 2:1 type, with a centrosymmetric crystal structure (monoclinic–prismatic class, space group $C\,2/m$). This mineral has proved effective in catalyzing the formation of oligomers of RNA containing up to 30–50 monomer units.[25] This catalyst controls the structure of the oligomers formed, restricting the generation of highly complex isomeric mixtures. Montmorillonite-catalyzed reactions of 5′-activated D,L-adenosine and D,L-uridine yielded oligomers which extended to heptamers and hexamers, respectively. With these racemic monomers a modest enrichment in the formation of homochiral products was observed. For 5′-activated D,L-adenosine, the ratio of the proportion of homochiral products to the expected statistical distribution was 1.3, 1.6, and 2.1, for the dimers, trimers and tetramers, respectively. For 5′-activated D,L-uridine, homochiral products did not predominate for small oligomers, with ratios of dimers, trimers and tetramers 0.8, 0.44, and 1.4, respectively.[26] In either case, however, the homochiral predominance seemed to be enhanced with chain length.

It can be inferred that selective adsorption of linear arrays of D- and L-amino acids (or other relevant prebiotic molecules) on the surface of minerals, followed by condensation and polymerization, offers a plausible mechanism for the production of homochiral polypeptides or other biopolymers in the prebiotic Earth. Although some of these chiral surfaces can serve as effective asymmetric inductors, however, they do not in themselves provide the answer to the problem. It is no surprise that surface chemistry is a powerful aid, and that it might have been of great relevance in the prebiotic synthetic chemistry— but it is unclear how chiral catalytic activity in a given sense could prevail over its equally probable mirror image at the same time in a different location. In any local deterministic scenario, if the area of study is large enough, the overall chiral effect cancels out (*i.e.*, it is racemic) due to the equal and random occurrence of each handedness in mineral matter.

8.2.3 Organic Crystals: Glycine

Chiral faces may also be found on organic crystals. Glycine is the simplest and most abundant amino acid found in meteorites (Section 7.2, *Meteorites*) and its presence has been detected a number of times in the interstellar medium.[27] Recently radio astronomers have for the first time detected its precursors ($H_2C=NH$ and HCN) outside the Milky Way, in the ultra-bright galaxy Arp 220.[28] α-Glycine is one of the polymorphs of the amino acid glycine in the solid state, and crystallizes in a centrosymmetric structure (space group $P\,2_1/n$).

Crystals of α-glycine floating at the air–water interface expose only one enantiomorphic face towards the solution. Experiments with a number of racemic α-amino acids show that (*R*)-enantiomers are preferentially adsorbed at the (010) face of the α-glycine crystals, and the (*S*)-enantiomers at the (0–10) face. When solutions containing different racemic α-amino acids and glycine are evaporated, the first glycine crystal formed by chance will preferentially adsorb only the corresponding enantiomer from the racemic mixture, resulting in an enantioenriched solution of the remaining amino acid. If this initial chiral imbalance in the solution is capable of inducing the crystallization of glycine with the same face as in the initial crystal oriented towards the solution (and this is certainly an unexpected outcome), a self-amplificating process will occur, resulting in the enrichment of the less preferred adsorbed enantiomer in the solution, and hence enantiomeric resolution.[29]

8.3 Two-Dimensional Chirality

The boundary between two immiscible homogeneous phases, such as gas–liquid, liquid–liquid or liquid–solid, constitutes in itself a restriction in the degrees of freedom of any molecule located on it when compared to a molecule in the bulk homogeneous phase. This situation entails the preferential ordering of the molecules between the two phases, or self-assemblies, governed by the most favorable balance of certain directional energetic interactions with the phases, and which tends to be disrupted by the effect of temperature. These directional interactions include solvation and hydrophobic effects, amongst others.

A striking example of this type of arrangement is provided by the self-assemblies formed at the air–water surface. Amino acids functionalized with long aliphatic chains derived from fatty acids (*e.g.*, stearic acid, $C_{17}H_{35}CO_2H$) behave as amphiphilic molecules (*i.e.*, possessing both hydrophilic and hydrophobic properties). At the water–air interface these molecules form two dimensional (2D) self-assemblies in which the polar amino acid ends are solvated by the aqueous phase and the hydrophobic alkyl chains arrange themselves above the water surface, in the same way as surfactants. The ordering is profound, displaying network regularities which can be determined by GIRX (grazing incidence X-ray diffraction). Different racemic mixtures of amino acids functionalized with long aliphatic chains give different self-assemblies. These so-called 2D crystallites occur in three forms:

- racemic compounds, in which both enantiomers are packed together in the crystallite;
- conglomerates, involving the segregation of enantiomers, *i.e.*, crystallites are present showing each type of handedness; and
- disordered solutions, with random distribution of molecules of both handedness.

This nomenclature evokes the terminology generally used to describe the different 3D crystalline phases of actual racemic mixtures (Section 5.6.1, *Solubility Properties*). The analogies are clear. Under suitable conditions, these 2D crystallites can be polymerized, giving stereocontrolled peptides with homochiral, heterochiral or random sequences (Figure 8.4A and B). The actual stereochemical channel adopted during polymerization depends upon the exact chemical make-up of the amino acids. Derived from lysine, the N^ε-stearyl lysine thioethyl ester (C_{18}-TE-Lys) is a representative example of an amino acid derivative forming racemic 2D crystallites (Figure 8.4B). Polymerization of these racemic crystallites occurs in a stereoselective manner, producing homochiral peptides, but necessarily in the form of a racemic mixture (Figure 8.4B, *right*: isotactic polymerization).

In line with this two-dimensional architecture, the role that might have been played by ordered 2D assemblies of amino acid derivatives at the air–water interface in the generation of enantiopure homochiral oligopeptides has been studied. These 2D crystallites can indeed promote chiral amplification if there is initially an enantiomeric imbalance of the amino acid derivative (or scalemic mixture). The mechanism implies self-assembly of the L- and D-amino acid components of the scalemic mixture into racemic crystallites until the minor enantiomer has been exhausted, as in Figure 8.4B. The excess of the remaining major enantiomer will form homochiral domains, as in Figure 8.4A.

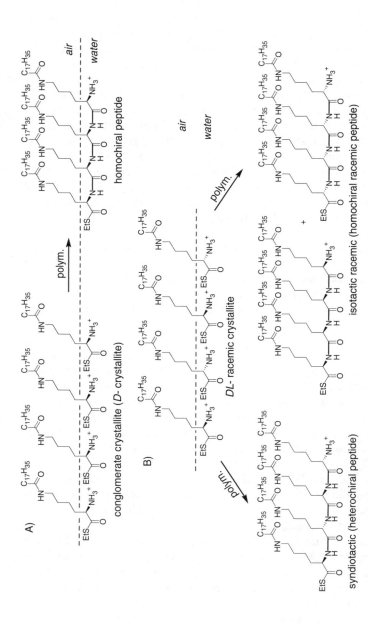

Figure 8.4 The activated lysine derivative C$_{18}$-TE-Lys (N^ε-stearyl lysine thioethyl ester) is an amphiphilic molecule which forms 2D crystallites at the air–water interface. These self-assemblies can be polymerized, yielding oligopeptides in a stereoselective manner. The theoretically possible outcomes are as follows: (*A*) If it forms a 2D conglomerate, each homochiral domain (or crystallite) polymerizes to give homochiral peptides; only the enantiopure crystallite derived from D-lysine is shown, but there are an equal number of L-lysine crystallites. (*B*) If it forms a racemic crystallite, two stereoselective polymerizations of syndiotactic stereochemistry (alternating D- and L-monomers). On the *left*, the 2D racemic crystallite may form a heterochiral peptide of syndiotactic stereochemistry (alternating D- and L-monomers). On the *right*, the 2D racemic crystallite produces racemic mixtures of homochiral peptides on polymerization. This last type of polymerization is the channel actually adopted for C$_{18}$-TE-Lys, although examples of each kind are possible for different amino acid derivatives.

If the polymerization of the racemic crystallite takes place between molecules of opposite handedness (as in Figure 8.4, *left*), *i.e.*, producing syndiotactic polymers, the resulting mixture of polymers will be enantiomerically enriched, or even enantiopure, in the homochiral peptide. The homochiral peptide arises from the polymerization of the molecules making up the homochiral crystallite. The glutamic acid derivative, γ-stearylglutamic acid-*N*-carboxyanhydride (C_{18}-Glu-NCA), is a representative example of molecules behaving in this manner.[30]

To summarize, this interesting approach using 2D chemistry allows the generation of racemic homochiral oligopeptides from racemic mixtures of activated α-amino acids, or the generation of homochiral oligopeptides of single-handedness from chiral nonracemic mixtures of monomers.[31] The extensive work which accumulated on enantiomeric cross-inhibition between monomers during the formation of polymeric structures (*e.g.*, oligonucleotides) led to the erroneous idea that long-chain polymerization could only be achieved if nonracemic monomers were employed. Even when directed by a homochiral template, polymerization was severely cross-inhibited by the presence of the opposite enantiomer.[32] This is true in the ambit of polymerizations in solution, but efficient organization of the otherwise randomly oriented molecules in the bulk of the solution has been demonstrated to take place in such 2D crystallites. The application of these techniques to the preparations of homochiral peptides of heterogeneous composition might provide very valuable information.

It is likely that only long-chain homochiral polymers possess the degree of complexity necessary for the development of the primeval biological functions of life. An important corollary of this 2D chemistry is that the prerequisite of initial enantiomeric purity, once thought to be necessary to form homochiral macromolecules, no longer applies. A potential route to large homochiral macromolecules endowed with a high level of complexity, and therefore functional and evolutionary attributes, is available. The repercussions of this work towards the development of biotic theories is clear.

References

1. (a) W. A. Bonner, P. R. Kavasmaneck, F. S. Martin and J. J. Flores, *Science*, 1974, **186**, 143–144; (b) W. A. Bonner and P. R. Kavasmaneck, *J. Org. Chem.*, 1976, **41**, 2225–2226; (c) P. R. Kavasmaneck and W. A. Bonner, *J. Am. Chem. Soc.*, 1977, **99**, 44–50.
2. R. M. Hazen, T. R. Filley and G. A. Goodfriend, *Proc. Natl. Acad. Sci. USA*, 2001, **98**, 5487–5490.
3. A. M. Cody and R. D. Cody, *J Cryst. Growth*, 1991, **113**, 508–519.
4. P. Cintas, *Angew. Chem., Int. Ed..*, 2002, **41**, 1139–1145.
5. A. F. Wells, *Structural Inorganic Chemistry*, Oxford University Press, New York, 5th edn., 1984.
6. J. R. Chelikowsky, N. Troullier and J. L. Martins, *Phys. Rev. B*, 1991, **44**, 489–497.

7. J. Glinnemann, H. E. King Jr., H. Schulz, T. Hahn, S. J. la Placa and F. Dacol, *Z. Kristallogr.*, 1992, **198**, 177–212.
8. E. S. Dana, *A Textbook of Mineralogy*, Wiley, New York, 1949.
9. (a) S. F. Mason, *Nature*, 1984, **311**, 19–23; (b) G. E. Tranter, *Biosystems*, 1987, **20**, 37–48.
10. E. I. Klabunovskii, *Astrobiol.*, 2001, **1**, 127–131.
11. An excellent illustrative monograph on the mineral quartz is available at http://www.quartzpage.de.
12. H. Churchill, H. Teng and R. M. Hazen, *Am. Mineral.*, 2004, **89**, 1048–1055.
13. (a) J. D. Bernal, *Proc. Phys. Soc.*, 1949, **62**, 537–543; (b) M. Paecht-Horowitz, J. Berger and A. Katchalsky, *Nature*, 1970, **228**, 636–638; (c) A. G. Cairns-Smith, *New Scientist*, 1974, **61**, 274–276; (c) N. Lahav, D. White and S. Chang, *Science*,1978, **201**, 67–69; (d) L. M. Coyne, J. Lawless, N. Lahav, S. Sutton and M. Sweenev, *Origins of Life*, 1981, **11**, 115–124; (e) A. Yamagishi, *J. Chem. Soc., Dalton Trans.*, 1983, 679–681.
14. (a) M. M. Mortland, *J. Mol. Cat.*, 1984, **27**, 143–155; (b) A. Naidja and B. Siffert, *Clay Miner.*, 1989, **24**, 649–661.
15. (a) E. T. Degens, J. Matheja and T. A. Jackson, Nature, 1970, **227**, 492–493; (b) T. A. Jackson, *Experiencia*, 1971, **27**, 242–244.
16. (a) W. A. Bonner and J. Flores, *Curr. Mod. Biol.*, 1973, **5**, 103–113; (b) J. Flores and W. A. Bonner, *J. Mol. Evol.*, 1974, **3**, 49–56; (c) J. J. McCullogh and R. Lemmon, *J. Mol. Evol.*, 1974, **3**, 57–61.
17. (a) S. C. Bondy and M. E. Harrington, *Science*, 1979, **203**, 1243–1244; (b) B. Siffert and A. Naidja, *Clay Miner.*, 1992, **27**, 109–118; (c) H. Hashizume, B. K. G. Theng and A. Yamagishi, *Clay Miner.*, 2002, **37**, 551–557.
18. J. B. Youatt and R. D. Brown, *Science*, 1981, **212**, 1145–1146.
19. S. W. Bailey, in *Crystal Structures of Clay Minerals and their X-ray Identification*, G. W. Brindley and G. Brown ed., Mineralogical Society, London, 1980, pp. 1–124.
20. H. L. Son, Y. Suwannachot, J. Bujdak and B. M. Rode, *Inorg. Chim. Acta*, 1998, **272**, 89–94.
21. J. G. Thompson and C. Cuff, *Clays Clay Miner.*, 1985, **33**, 490–500.
22. A. Julg, A. Favier and Y. Ozias, *Struct. Chem.*, 1990, **1**, 137–141.
23. A. Julg, *Comp. Rend. Acad. Sci. II*, 1987, **305**, 563–565.
24. A. Julg, *THEOCHEM.*, 1989, **53**, 131–142.
25. (a) W. Huang and J. P. Ferris, *J. Am. Chem. Soc.*, 2006, **128**, 8914–8919; (b) W. Huang and J. P. Ferris, *Chem. Comm.*, 2003, **9**, 1458–1459.
26. P. C. Joshi, S. Pitsch and J. P. Ferris, *Orig. Life Evol. Biosph.*, 2007, **37**, 3–26.
27. L. E. Snyder, F. J. Lovas, J. M. Hollis, D. N. Friedel, P. R. Jewell, A. Remijan, V. V. Ilyushin, E. A. Alekseev and S. F. Dyubko, *Astrophys. J.*, 2005, **619**, 914–930.

28. Preliminary report: http://www.astrobio.net/news/article2623.html.
29. I. Weissbuch, L. Addadi, L. Leiserowitz and M. Lahav, *J. Am. Chem. Soc.*, 1988, **110**, 561–567.
30. H. Zepik, E. Shavit, M. Tang, T. R. Jensen, K. Kjaer, G. Bolbach, L. Leiserowitz, I. Weissbuch and M. Lahav, *Science*, 2002, **295**, 1266–1369.
31. I. Weissbuch, G. Bolbach, L. Leiserowitz and M. Lahav, *Orig. Life Evol. Biosph.*, 2004, **34**, 79–92.
32. (a) J. G. Schmidt, P. E. Nielsen and L. E. Orgel, *J. Am. Chem. Soc.*, 1997, **119**, 1494–1495; (b) G. F. Joyce, G. M. Visser, C. A. A. van Boeckel, J. H. van Boom, L. E. Orgel and J. van Westrenen, *Nature*, 1984, **310**, 602–604; (c) I. A. Kozlov, P. K. Politis, S. Pitsch, P. Herdewijn and L. E. Orgel, *J. Am. Chem. Soc.*, 1999, **121**, 1108–1109.

CHAPTER 9
Intrinsic Asymmetry of the Universe: The Arrow of Space–Time and the Unequal Occurrence of Matter and Antimatter

In the traditional view of the origin of the Universe, cosmologists believe that matter and antimatter were created during the early Big Bang from pure energy, in a perfectly symmetrical state. Such a situation of pure energy led to the creation of a particle–antiparticle combination on a very large scale. In the first instants following the Big Bang the energy density of the Universe was so high that photons (and other force carriers) started to interact in large numbers and formed quark–antiquark pairs, the particles of matter and antimatter (Section 4.2.1, *The Fundamental Interactions: Gravitational, Weak Interaction, Electromagnetic and Strong Interaction*). At that time the Universe underwent its first "phase transition" and gravity separated first from the other three forces, which could not be distinguished from one another at the temperatures involved. Later the strong force separated from the electroweak force and in further phase transitions occurring under cooler conditions led to the individualized fundamental forces (Section 4.2.4, *Unification of Forces*). In a projection of that initial symmetry, the Big Bang is thought to have created equal amounts of matter and antimatter, but according to best current knowledge all the antimatter has disappeared—along with most of the matter.

If particles of matter and antimatter were the exact opposites of each other they should have become mutually annihilated to leave only photons. However, the existence of our matter-dominated Universe suggests that matter and antimatter underwent different processes after the Big Bang. Somehow a tiny

---

The Origin of Chirality in the Molecules of Life
Albert Guijarro and Miguel Yus
© Albert Guijarro and Miguel Yus, 2009
Published by the Royal Society of Chemistry, www.rsc.org

amount of matter, about one matter particle in 10^9, survived. This is the photon to baryon number, $N_\gamma/N_B = 10^9$, obtainable from the intensity of the 3 Kelvin background radiation (see below). Antiquarks were annihilated by quarks, leaving behind only surplus quarks of normal matter. Quarks do not exist independently, so the surviving quarks bound themselves together into protons and neutrons (baryons) by means of the strong force. About 1 second after the Big Bang electron and positron pairs would have also been annihilated, leaving an excess of electrons.

Once the Universe was a few seconds old, it became cool enough for protons and neutrons to combine to form the first atomic nuclei, hydrogen and helium. By about 300 000 years, the Universe had become dilute and cool enough for light to escape. At this point, the temperature was about 3000 K, the electrons "recombined" with nuclei—strictly they actually combined for the first time, and the photons scattered off the now neutral atoms through a process named decoupling, and began to travel freely through space. The Universe had become "transparent". With the expansion of the Universe, the wavelength of these photons also stretched, and have continued stretching and cooling ever since; they have now reached 3 Kelvin temperature. This is the so called 3 K background radiation, a uniform radiation in the microwave region of the spectrum which shows the spectrum of a black body radiating at about 3 K, observed now in all directions in the sky. It is considered to be the remnant of the radiation emitted at the time the expanding universe became transparent.

Further expansion and cooling created the Universe as we know it, with matter and energy ruled by the four fundamental forces—and virtually no antimatter. Compelling evidence of this comes from observation of the cosmos. If regions of antimatter existed within our local cluster of galaxies we would be able to see the radiation caused by matter–antimatter annihilation at the boundaries. In addition, the cosmic microwave background radiation shows no evidence of disturbance from radiation due to subsequent annihilation, suggesting that there are no large regions of antimatter closer than at least 10^{10} light years, and possibly throughout the whole of the visible Universe.

To account for this excess of matter, the Standard Model of particle physics formulated in the 1970s can accommodate the fact that certain bizarre particles of matter and their antiparticles decay at slightly different rates. Known as charge parity ($\hat{C}\hat{P}$) violation, this strange phenomenon is one step in advance of parity ($\hat{P}$) violation (Section 4.2.3, *Parity Violation*), and was first indirectly observed in K mesons (or Kaons) by J. Cronin and V. Fitch in 1964 (see also Section 4.2.3.2, π^+ *and* π^- *Decay and the Conservation of the Symmetry Operator* $\hat{C}\hat{P}$).[1] In their experiments symmetry violation of the $\hat{C}\hat{P}$ transformation was observed for the neutral K_L^0 meson, which produces two sets of particles $K_L^0 \to \pi^- e^+ \nu$ and antiparticles $\bar{K}_L^0 \to \pi^+ e^- \bar{\nu}$ with a fractional excess of 3×10^{-3}. The effect is very tiny, which explains why it was not detected sooner. But it was consistently $\neq 0$ and hence indicative of $\hat{C}\hat{P}$ violation—and it merited a Nobel Prize.

It was then that the theorist A. Sakharov provided the key to answering the question on the origin of this asymmetry. In 1967 he used $\hat{C}\hat{P}$ violation to propose how the present matter-dominated Universe, which is in a far from thermodynamic equilibrium state, could have emerged from a condition containing exactly equal amounts of matter and antimatter in the earliest moments after the Big Bang.[2]

For many years the rare K_L^0 meson event mentioned above was the only experimental testimony for $\hat{C}\hat{P}$ violation. The term "meson" is used to refer to any particle made up of one quark (any "flavor", in the sense of any of the six "kinds" of quarks, see Figure 4.1) and one antiquark (also of any flavor). The quark and antiquark are mainly bound together by the strong force, and decay by changing flavor through weak force interactions. They probably represent the closest we can get to the quarks for the moment. Because quarks can change flavor by weak interactions, all heavier quarks decay to one or other of the lighter generations by mean of the weak force and only the lightest quarks (and leptons) are included in the stable matter. Within this context, in a simplified fashion the K^0 meson can be described as a "down-antistrange" ($d\bar{s}$) combination of quarks which decays to a generation of lighter quarks ($\pi^- = \bar{u}d$) and leptons ($e^+ v$).

The Standard Model also predicts an asymmetry in the decay rates of other esoteric mesons, known as B mesons. The B meson is an unstable particle consisting of a b-antiquark (called a bottom-antiquark, $\bar{b}$ or "b-bar") and either a u- or d-quark (up or down; these are the two lightest quarks). Its corresponding antiparticle, called the B antimeson, $\bar{B}$, or "B-bar" meson, is made up of a b-quark and a u- or a d-antiquark (u-bar or d-bar). It has a mass almost six times that of the proton, accounted for mainly by the mass of the bottom-quark it contains, which is almost as massive. In 2001, after 37 years of searching for further examples and further confirmation of $\hat{C}\hat{P}$ violation, the BaBar team at Stanford in the USA,[3] and independently by the Belle Collaboration in Tsukuba, Japan,[4] discovered indirect $\hat{C}\hat{P}$ violation in B mesons. By measuring the decay rates with extreme accuracy these groups were able to show that B mesons decay slightly more slowly, $B_0 \to \pi^- \pi^+$, than their antiparticle equivalents, $\bar{B}_0$.

It took until 2004 for BaBar scientists to observe direct evidence of the phenomenon. The most dramatic distinction between matter and antimatter to date was established by analyzing the decay of more than 200 million pairs of B and anti-B mesons. Among the many ways in which mesons can decay, the team were looking for rare events which convert B mesons and anti-B mesons into kaons–pions pairs, $B_0 \to K^+ \pi^-$ and $\bar{B}_0 \to K^- \pi^+$, respectively. The experiment showed a 13% higher decay rate for the particles than for their equivalent antiparticles, the largest difference so far observed.[5]

Physicists now accept that there are at least two kinds of subatomic particles which exhibit this puzzling phenomenon (Figure 9.1). $\hat{C}\hat{P}$ violation does not imply that matter and antimatter are necessarily different. We are speaking only about the *rates* at which particles decay. Provided that $\hat{C}\hat{P}\hat{T}$ is conserved, matter and antimatter remain degenerate and $\hat{C}\hat{P}$ can only give an excess of matter when conditions are far from equilibrium, as in the rapidly expanding Universe of Sakharov. This provides an analogy with true and false chirality, in which the latter could generate asymmetry only under kinetic conditions, while

Intrinsic Asymmetry of the Universe 141

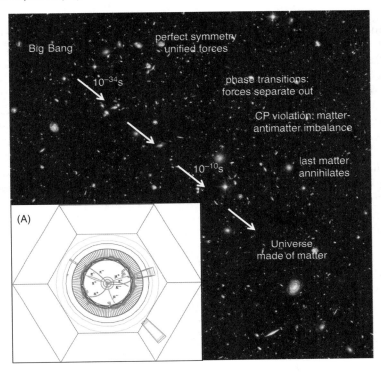

Figure 9.1 Antimatter is almost completely absent from today's Universe, being detected in only the tiniest quantities for fleeting moments in collisions between cosmic particles, and in accelerators. This overabundance of matter is attributed to a major symmetry breaking—the charge parity ($\hat{C}\hat{P}$) violation—which appeared about 10^{-34} s after the Big Bang. At around that time the Universe had suffered its first phase transition and the forces began to separate. The left-handed nature of the weak force is believed to be responsible for the $\hat{C}\hat{P}$ violation, shown by a tiny difference in the decay rates between particles of matter and antimatter. Once particle–antiparticle annihilation was completed 10^{-10} s after the Big Bang, this process left a small excess of matter—the matter from which our present Universe is constructed. In the picture (A) is shown a computer reconstruction of a "golden event" inside the BaBar detector. From the energy released by the annihilation of electrons and positrons in the centre of the detector, pairs B and anti-B are created, which typically decay into a spray of particles (muons, kaons and pions). The difference in decay rates between B and anti-B particles confirms the $\hat{C}\hat{P}$-violating nature of certain types of particle decay, such as that of the B meson. (Pictures are from the Hubble website (Hubble ultra-deep field image of the Universe): http://hubblesite.org/newscenter/newsdesk/archive/releases/2004/07/image/m, and (A) is shown by courtesy of Stanford Linear Accelerator Center: http://www.slac.stanford.edu/slac/media-info/photos/lg-event.html.

the two states remain degenerate at thermodynamic equilibrium. According to current knowledge of the phenomenon, $\hat{C}\hat{P}$ violation is thought to be responsible for the great preponderance of matter in our expanding Universe, while $\hat{C}\hat{P}\hat{T}$ is thought to be preserved. The existence of $\hat{C}\hat{P}$ violation is important not just because of its implications for symmetry. $\hat{C}\hat{P}$ violation entails $\hat{T}$ violation, and a number of fundamental concepts, including the mistaken conviction that all physical laws would apply equally well with time running forward or backward, *i.e.*, the microreversibility principle, do not apply here (Section 4.2.3.3, *The Principle of $\hat{C}\hat{P}\hat{T}$ Invariance*). This is what in particle physics is called the arrow of time, which deals with the possibility of time non-invariance ($\hat{T}$), and which should not be mistaken for the distinctly different and unrelated thermodynamic arrow of time. Different unconnected aspects of physics are involved here, namely the equations of motion of the involving particles for the former, and the second law of thermodynamics for the latter, which rests on statistics and its reliability arises from the huge number of particles present in macroscopic systems.

The occurrence of this fundamental asymmetry ($\hat{C}\hat{P}$ violation) immediately suggests the existence of a general arrow of space–time, which includes the possibility of actually discerning the sign of certain properties once considered non-observable, such as the absolute direction of time flow (from the past to the future, or *vice versa*), and the absolute handedness of chirality (right or left), which is of the greatest interest to us—the absolute sign of electric charge (positive or negative) would be equally defined along with the former. The existence or denial of a connection between absolute chirality and the chirality of the molecules of life has long been debated, but has still to be properly addressed. The efforts made so far in one or the opposite direction have been summarized and discussed in this book.

The Standard Model is able to accommodate the theoretical background to $\hat{C}\hat{P}$ violation in quarks, since:

- there are at least three generations of quarks, as predicted by M. Kobayashi and T. Maskawa in 1973,[6] all six quarks now having been identified;[7] and
- they interact *via* the weak force, a force that differentiates between left and right.

However, our current knowledge of $\hat{C}\hat{P}$ violation is still incomplete. The K_L^0 and B_0 examples are insufficient by many orders of magnitude to account for the amount of matter we can observe. Models involving heavy objects, for example models with additional $\hat{C}\hat{P}$-violating phases in the mass mixing of Higgs bosons or other undiscovered particles, could account for the observed excess of protons. Actual accelerator experiments, such as the Large Hadron Collider (LHC) at CERN,[8] projected to be fully operative during 2008, are aimed at discovering new sources of $\hat{C}\hat{P}$ violation and extending our knowledge of those already known.[9] Later this decade many top quarks—the heaviest one—and other exotic particles will be generated by the LHC, offering an opportunity to search directly for $\hat{C}\hat{P}$ violation associated with such very heavy states. At the energies involved the LHC is expected to reproduce the conditions operating about 10^{-10} s after the Big Bang, the time at which the electroweak force separated and the matter phase is still described in terms of a quark–gluon plasma.

Future colliders might go even further, covering the interval only 10^{-34} seconds after the Big Bang. This is described as the "Grand Unification" epoch, during which it is believed that the imbalance between particles and antiparticles arose. This epoch is described by the Grand Unified theories, which unite strong and electroweak forces. The X boson is a heavy particle predicted by this theory, which could have been abundant at that time, allowing quark–lepton interconversion. Inequality in the decay rates of the hypothetical X and $\bar{X}$ boson, as was the case with the B_0-meson, could have generated an excess of matter in the early Universe, described by Equations (9.1), (9.2) and (9.3). According to this theory a great deal of matter, and equal amounts of antimatter, was produced by pair creation from energy, Equation (1). Next, the X boson decayed into two quarks, or an antilepton and an antiquark, producing mainly matter, Equation (9.2). But there are an equal number of $\bar{X}$ bosons, which decay predominantly into antimatter in a charge-conjugated process, Equation (9.3). If the rate x = x' there is no $\hat{C}\hat{P}$ violation and the two decay branches are identical for the particle and antiparticle. Eventually, annihilation of the two will leave no trace of matter, only photons. But if $x > x'$ after the bulk of quark and lepton annihilations $q\bar{q}, l\bar{l} \to \gamma$, there would be a small excess of quarks left, equal to $3(x-x')$, forming a small number $(x-x')$ of baryons made up of three quarks (e.g. proton=uud) along with an equally small number $(x-x')$ of leptons, (e.g. electron) which eventually would lead to an excess of matter.[10]

$$\gamma \to X \bar{X} \tag{9.1}$$

$$\bar{q}\bar{l} \overset{1-x}{\longleftarrow} X \overset{x}{\longrightarrow} qq \tag{9.2}$$

$$q l \overset{1-x'}{\longleftarrow} \bar{X} \overset{x'}{\longrightarrow} \bar{q}\bar{q} \tag{9.3}$$

At present the existence of the X boson can not be verified directly, due to its mass. It is estimated to be 10^{14} times the mass of the proton, which is far beyond the reach of today's particle accelerators. It is still only a hypothetical model.

Nevertheless, $\hat{C}\hat{P}$ violation may appear in many different ways, mainly still unexplored.[11] It may occur in heavy quarks, or also in neutrinos.[12] Its source may lie in the properties of the Higgs boson,[13] or beyond the Standard Model,[14] in supersymmetry,[15] technicolor,[16] or even in extra dimensions in the realm of superstrings,[17] a theory which has captivated an increasing number of physicists in recent years due to the sheer beauty of its mathematics and its ability to resolve difficult problems. The LHC will provide the opportunity to test many of these concepts and theories (Figure 9.2). It is possible that a new "Theory of Everything" may arise, along with new

Figure 9.2 The Large Hadron Collider (LHC), currently in the final stages of construction at CERN and expected to become operative during 2008, will allow proton–antiproton beams to collide at higher energies than ever before. Many experiments in the LHC will study matter–antimatter asymmetry and related phenomena. The LHC-B will pick out collisions of particles bearing the b-quark (like B_0 mesons) and its CP-violating properties. ATLAS is a huge new detector that will search for extra dimensions and CMS will explore the hypothetical Higgs particle. In ALICE, the LHC will attempt to study the formation of the quark–gluon plasma, a state of matter that could have existed early after the Big Bang. Two experiments, ATHENA and ATRAP, have already reported the creation of large quantities of anti-hydrogen atoms (around 100 000 each) and will look for spectroscopic differences and test $\hat{C}\hat{P}\hat{T}$ symmetry and current parity violation theory. In place of the now obsolete LEP (Large Electron Positron Collider), the LHC will be the most powerful instrument ever built to investigate the properties of matter, including its inherently asymmetric nature. (Pictures from CERN: http://outreach.web.cern.ch/outreach/public/cern/PicturePacks/Picturepacks.html.)

answers. In any case exciting times lie ahead in fundamental physics, and the newest technologies will be needed to test the theories which emerge. The repercussions are likely to be far reaching, perhaps extending into new territories.

References

1. J. H. Christenson, J. W. Cronin, V. L. Fitch and R. Turlay, *Phys. Rev. Lett.*, 1964, **13**, 138–140.
2. (a) A. D. Sakharov, *Pis. Zh. Eksp. Teor. Fiz.*, 1967, **5**, 32–35; (b) A. D. Sakharov, *JETP Lett.*, 1967, **5**, 24–27.
3. (a) BaBar Collaboration, B. Aubert, et al., *Phys. Rev. Lett.*, 2001, **87**, 091801/8; (b) BaBar Collaboration, B. Aubert, et al., *Phys. Rev. D*, 2002, **66**, 032003/54; (c) BaBar Collaboration, B. Aubert, et al., *Phys. Rev. Lett.*, 2002, **89**, 201802/7.
4. (a) Belle Collaboration, K. Abe, et al., *Phys. Rev. Lett.*, 2001, **87**, 091802/7; (b) Belle Collaboration, K. Abe, et al., *Phys. Rev. D*, 2002, **66**, 032007/22.
5. BaBar Collaboration, B. Aubert, et al., *Phys. Rev. Lett.*, 2004, **93**, 131801/7.
6. M. Kobayashi and T. Maskawa, *Prog. Theor. Phys.*, 1973, **49**, 652–657.
7. (a) CDF Collaboration, F. Abe, et al., *Phys. Rev. Lett.*, 1995, **74**, 2626–2631; (b) DØ Collaboration, S. Abachi, et al., *Phys. Rev. Lett.*, 1995, **74**, 2632–2637.
8. http://lhc-new-homepage.web.cern.ch.
9. T. Nakada, *Eur. Phys. J. C.*, 2004, **34**, s49–s53.
10. A. J. MacDermott, in *Chirality in Natural and Applied Science*, W. J. Lough and I. W. Wainer ed., Blackwell Science, Oxford, 2002, 46–48.
11. I. I. Bigi and A. Sanda, *CP violation*, Cambridge University Press, Cambridge, 1999, 269–358.
12. (a) P. H. Frampton, S. L. Glashow and T. Yanagida, *Phys. Lett. B*, 2002, **548**, 119, 121; (b) P. Bargueño and R. P. de Tudela, *Orig. Life Evol. Biosph.*, 2007, **37**, 253–257.
13. P. Renton, *Nature*, 2004, **428**, 141–144.
14. (a) J. L. Rosner, *Am. J. Phys.*, 2003, **71**, 302–318; (b) Y. Nir, http://arxiv.org/abs/hep-ph/0109090v1.
15. A. Masiero and O. Vives, *Nucl. Phys. B*, 2001, **101**, 253–262.
16. A. Martin and K. Lane, *Phys. Rev. D*, 2005, **71**, 015011/24.
17. (a) D. A. Demir and L. L. Everett, *Phys. Rev. D*, 2004, **69**, 015008/17; (b) M. Brhlik, L. Everett, G. L. Kane and J. Lykken, *Phys. Rev. D*, 2000, **62**, 035005/14.

Subject Index

abiotic theories 7
achiral molecules 48
aggregates 85
amino acids 43
　asymmetric photolysis by stellar CPL 56
　eutectic ee values 96
　in meteorites 116–18, 119
　N-carboxyanhydrides 86
amplification 72–107
　autocatalysis see autocatalysis
　eutectic mixtures 91–8
　Frank model 72–5, 109, 111
　nonlinear effects 78–9, 84–6
　in polymerizations 86–7
　Salam phase transition 91
　self-replication 82–4
　in serine octamers 98–102
　supramolecular assemblies 87–8
　of weak force 77–8
　Yamagata cumulative mechanism 88–91
anisotropy 49, 59
antimatter, fate of 138–9, 141
antineutrinos 34, 35, 37
antiparticles 32
antiquarks 139, 140
Arago, F. 3
arrow of time 142
asymmetric autocatalysis 78–82
atomic force microscopy 63
autocatalysis 72–84
　amplification of weak force 77–8

　asymmetric 78–82
　closed system 75
　open-flow system 75, 76, 77
　see also amplification
axial (pseudo) vectors 23

3 K background radiation 139
Barron, L. D. 22
baryons 32, 139
Bernoulli trials 8
beta-decay 34–5, 46–8
Big Bang theory 138
1,1'-binaphthyl 74, 110
binary phase diagrams 93–4
biomolecular homochirality 6–20
　theories of 7
biotic theories 7
Biot, J.-B. 3
Bose–Einstein condensation 91
bosons
　X-type 143
　neutral $Z°$-type 39–41
　weak 34
Bremsstrahlung radiation 47
bromochlorofluoromethane 45, 46

calcite 129–32
carbonaceous chondrites 56, 118, 120
chance mechanisms 6, 7–17, 115
　amplification of stochastic imbalances 8–10
　macromolecules 10–17
charge conjugation 24–5

Subject Index

charge parity violation 35–8, 139, 142
  conservation of symmetry operator 37–8
  CPT invariance 38
  energy shift 44–6
  experimental confirmation 35–7
  quantification of 41–6
chiral crystals 125–32
  calcite 129–32
  glycine 132
  gypsum 129–32
  quartz 2–3, 90, 125–9
  see also individual crystals
chiral field, absence of 25
chirality 21–30
  false 22, 28–9
  true 2–7, 22
chiral transfer 65
chiroselectivity 84
chondrites, carbonaceous 56, 118, 120
chromium complexes 57–9
circular dichroism 48–57
circularly polarized light
  on Earth 52
  and handedness 56
  in outer space 52–7
clay minerals 129–32
closed systems 75
Cometary Sampling and Composition (COSAC) 122
comets 120–3
conglomerates 92
  solubility of 94
conservation of symmetry operator 37–8
Coriolis effect 65
cosmic microwave radiation 139
CPL *see* circularly polarized light
CPT invariance 38
Cronin, J. 139
crystallization, spontaneous symmetry breaking in 110–13
Curie, Pierre 25

decoupling 139
deterministic mechanisms 17–18, 115, 125–37

diastereomeric interactions 44
4,4′-dimethyl chalcone 110
Direc delta distribution 41
dissipative structures 74
dissolution 86
dopants 88

eigenvalues 42
electromagnetic interaction 31–4, 138
electroweak theory 39
enantiodiscrimination 7
enantiometers 1
enthalpy of solution 93
eutectic mixtures 91–8
  solubility properties 93–6
  sublimation properties 97–8

false chirality 22
  effect on molecules 28–9
Faraday effect 21, 52
Fermi weak coupling constant 41
Fitch, V. 139
fluid dynamics, vortex motion 59–65
Frank, F. C. 72
Frank model 72–5, 109, 111
  models derived from 75–7
Fresnel, A. J. 3

geothermal energy 47
Gibb's phase rule 94, 95
Glashow, S. 39
glycine crystals 132
Grand Unification epoch 143
gravitation interactions 31–4
gypsum 129–32

hadrons 32
Hajos–Parrish reaction 84
Halley's comet 122
halogenomethanes 45
Haüy, R. H. 2
helicenes 50
hemihedral facets 2
  see also chiral crystals
hemimorphism 127
Herschel, J. W. F. 3

hexahelicene, asymmetric
  photosynthesis 50–1
Higgs bosons 142
history 1–5
holohedral crystals 2
  see also chiral crystals
homochirality 75
  origins of 6–20
homochiral preference 98–9

initiators 88
isomorphism 3
Ivuna meteorite 118, 121

J-aggregates 61–4

Kagan, H. B. 84
kaolinite 131
Kaons 139
Kelvin, William Thomson, Lord 21
Kondepudi, D. K. 78
Kuhn–Condom sum rule 51–2, 56–7
Kuiper belt 120

Large Hadron Collider 142–3, 144
Lee, T. D. 35
leptons 32, 33, 41
linear dichroism 62
local mechanisms see deterministic
  mechanisms

macromolecules 10–17
magnetochiral birefringence 57–9
magnetochiral dichroism 26, 57–9
majority rules effect 87
matter 138
mesons 32, 37, 140
  B 140
  K 139
  decay rate asymmetry 140
mesophases 75, 110
meteorites 115–20, 121
  amino acids in 116–18, 119
  L-enantiomeric enrichment 117–18
  soluble organic compounds 116
mirror-symmetry breaking 78–82

Mitscherlich, E. 3
Monte Carlo simulation 43
montmorillonite 131–2
muons 37
Murchison meteorite 115–20, 121
Murray meteorite 121
mutual antagonism 73, 79

neutral $Z°$ boson 39–41
neutrinos 35, 37
neutrons 139
NGC 6334 55
nonlinear effects 78–9, 84–6
Nylon-1 87

oligonucleotides, self-replication 82–3
OMC-1 molecular cloud 54, 55
Oort cloud 120
open-flow systems 75, 76, 77
optical activity 2
optical rotation 3
  magnetically induced 21
organic crystals 132
  see also chiral crystals
Orgueil meteorite 118
Orion Nebula 53, 54
Ostwald ripening 112
outer space 115–24
  circularly polarized light in 52–7
  comets 120–3
  meteorites 115–20

panspermia 55
parity conservation 35
parity violation see charge parity
  violation
particle physics see Standard Model
Pasteur, Louis 1–2, 3, 21, 59
  failure of experiments 25–7
peptides, self-replicating 83
photolysis, asymmetric 48–59
photosynthesis, asymmetric 48–59
polar (true) vectors 23
polycyclic aromatic hydrocarbons 56
polymerizations 10–17
  amplification in 86–7

protons 139
pseudoracemates 92

quantum chromodynamics 32
quarks 32, 33, 41, 140, 142
quartz crystals 2–3, 90, 125–9
  hemimorphism 127
  internal structure 128–9
  twinned 127

racemic compounds 92
  solubility 94–5
racemic mixtures 1, 48
  composition of 9
racemic solid solutions 92
racemic state 8–10
racemization 56
Raupach, E. 57
replicators 83
Ribó, J. M. 61
Rikken, G. L. J. A. 57
RNA, polymeric chains 14–16
Rosetta mission 122–4
rotovibrational spectroscopy 45
Rubenstein–Bonner neutron star hypothesis 55

Sakharov, A. 139
Salam, A. 39, 91
Salam phase transition 91
salt-induced peptide formation reaction 84
scalemic mixtures 86, 91–8
self-replication 82–4
sergeants-and-soldiers effect 87
serine octamers 98–102
  chemistry 99–101
  homochiral preference 98–9
  sublimations 102
Soai, K. 74, 75, 78
sodium bromate 110
sodium chlorate 74, 110, 111
Solar System, formation of 55–7
space inversion 22–3, 36
spin–orbit coupling 41
spontaneous symmetry breaking 108–14
  in crystallization 110–13

Standard Model of particle physics 32, 33, 40, 139–40
stellar radiation 53
Stirling approximation 9, 109
stirring
  and chirality 59
  vortical 62, 63, 65
stochastic fluctuation 109
stochastic imbalance, amplication of 8–10
Strecker reaction 118
strong interaction 31–4
sublimation 86, 97–8
supramolecular assemblies 87–8
  achiral 88
  chiral 88
symmetry breaking 62, 63–4, 74
  spontaneous 108–14
symmetry operations 22–5
  charge conjugation 24–5
  space inversion 22–3
  time reversal 23–4

tartaric acid 1, 3–4
tartrates 3–4
ternary phase systems 95, 96
time non-invariance 142
time reversion 23–4
true chirality 2–7, 22
two-dimensional chirality 132–5

Ulbrich, T. L. V. 47
unification of forces 38–41
  electroweak theory 39
  neutral $Z°$ boson 39–41
Universe, formation of 138–9

vectors
  axial (pseudo) 23
  polar (true) 23
Vester, F. 47
vortex motion 59–65
  macroscopic level 65
vortical stirring 62, 63, 65

weak interaction 31–4
  amplification of 77–8

Weinberg angle 41
Weinberg, S. L. 39
Wigner, E. P. 35
Wu, C. S. 35

Yamagata cumulative mechanism 88–91
Yang, C. N. 35

zwitterions 43